Promoting Joint Warfighting Proficiency

The Role of Doctrine in Preparing Airmen for Joint Operations

Miranda Priebe, Laurinda L. Rohn, Alyssa Demus, Bruce McClintock, Derek Eaton, Sarah Harting, Maria McCollester

Prepared for the United States Air Force

For more information on this publication, visit www.rand.org/t/RR2472

Library of Congress Cataloging-in-Publication Data is available for this publication.
ISBN: 978-1-9774-0054-3

Published by the RAND Corporation, Santa Monica, Calif.

Preface

In September 2016, Chief of Staff of the Air Force Gen David Goldfein announced an initiative to strengthen joint leaders and teams. This initiative is looking, in part, at how to improve joint warfighting proficiency so airmen can better integrate into, influence, and lead joint teams. Differences between Air Force and joint doctrine could contribute to airmen speaking a different language than their joint counterparts or being unfamiliar with the processes and principles employed in a joint setting. The Air Force asked RAND to describe the current differences and similarities between Air Force and joint doctrine and identify ways to increase doctrinal alignment. This report also considers how airmen currently encounter joint doctrine to answer two questions. First, to what extent could greater alignment of Air Force and joint doctrine help airmen gain greater joint proficiency? Second, would changes to the way Air Force officers learn or use doctrine help to promote joint proficiency?

The research reported here was commissioned by Maj Gen Brian M. Killough, director of Strategic Plans, Office of the Deputy Chief of Staff for Strategic Plans and Requirements, Headquarters, U.S. Air Force, and conducted within the Strategy and Doctrine Program of RAND Project AIR FORCE as part of the project "Air Force and Joint Doctrine Gap Analysis."

RAND Project AIR FORCE

RAND Project AIR FORCE (PAF), a division of the RAND Corporation, is the U.S. Air Force's federally funded research and development center for studies and analyses. PAF provides the Air Force with independent analyses of policy alternatives affecting the development, employment, combat readiness, and support of current and future air, space, and cyber forces. Research is conducted in four programs: Strategy and Doctrine; Force Modernization and Employment; Manpower, Personnel, and Training; and Resource Management. The research reported here was prepared under contract FA7014-16-D-1000.

Additional information about PAF is available on our website: www.rand.org/paf/

This report documents work originally shared with the U.S. Air Force on November 15, 2017. The draft report, issued on December 12, 2017, was reviewed by formal peer reviewers and U.S. Air Force subject-matter experts.

Contents

Figures

Table

Summary

Air Force and joint operational concepts call for rapid integration of U.S. military capabilities across multiple domains.[1] Future joint operations, therefore, will require even greater cooperation among the services and the ability to integrate capabilities across domains at lower echelons of command. As a result, Chief of Staff of the Air Force Gen David Goldfein announced an initiative to increase airmen's joint warfighting proficiency, so they can better integrate into, influence, and lead current and future joint teams.

As part of the initiative, the Air Force asked RAND Corporation researchers to describe the differences and similarities between Air Force and joint doctrine and identify ways to increase doctrinal alignment. This report also considers how airmen currently encounter doctrine to answer two questions. First, to what extent could greater alignment of Air Force and joint doctrine help airmen gain greater joint proficiency? Second, would changes to the way Air Force officers learn or use doctrine help to promote joint proficiency?

A fundamental building block of joint proficiency is understanding joint doctrine. Joint doctrine captures the operational principles that prevail when the services go to war together. It also includes joint constructs (the terms, relationships, organizations, processes, and principles that the services share) and the joint mindset (a view that the services integrate, not just coordinate or deconflict when they fight as part of a joint team).

Although airmen do not have to use joint doctrine all the time to be effective in joint settings, using joint terminology and other constructs more regularly could enhance joint proficiency. In theory, ensuring that service doctrine is largely consistent with and provides linkages to joint doctrine is one way to give service members more exposure to these constructs earlier in their careers.[2] In practice, the impact of service doctrine on joint proficiency depends on how the service uses and values doctrine.

As a result, this report assesses both the alignment of Air Force and joint doctrine and the role of doctrine in the Air Force today. The doctrine comparison revealed that Air Force doctrine inconsistently uses joint constructs and sometimes has a service-centric tone. Because doctrine reflects a service's culture and extant practices, these findings suggest that acceptance of joint constructs and a joint mindset are also uneven. We also explored the role of operational doctrine within the Air Force and examined how airmen learn and use joint constructs. This analysis suggests

[1] U.S. Air Force, *Air Force Future Operating Concept: A View of the Air Force in 2035*, Washington, D.C., 2015, pp. 8–9; U.S. Department of Defense, *Joint Operational Access Concept (JOAC)*, Washington, D.C., January 17, 2012, pp. 17, 21; Michael E. Hutchens et al., "Joint Concept for Access and Maneuver in the Global Commons: A New Joint Operational Concept," *Joint Forces Quarterly*, Vol. 84, First Quarter 2017.

[2] Inconsistencies might sometimes be necessary because of the unique characteristics of a particular service. Therefore, explaining the differences and logic behind them is another part of promoting joint proficiency.

that doctrine is not widely valued in the Air Force. As a result, revisions to Air Force doctrine, on their own, would only have a limited impact. Providing airmen more opportunities to regularly use joint constructs throughout their careers may have a more significant impact on joint proficiency.

Building a Framework for Assessing Doctrinal Alignment

Chapter 2 of this report asks: Does Air Force doctrine use joint terms, promote joint processes and principles, and encourage an integrated joint mindset? As an initial step to answering this question, we developed an analytical framework for assessing the overall alignment between any service's doctrine and joint doctrine. First, we identified the key areas of doctrine for comparison: organization of doctrine; assumptions about operating environment and missions; operational principles, including the role of airpower in joint operations; roles, responsibilities, and processes; and terminology.

Second, we developed criteria for assessing the substantive alignment and tone of a service's doctrine. The substantive alignment of documents ranges from significantly divergent to significantly aligned with joint doctrine. Criteria for significant alignment are the following: Service principles, processes, and terms are consistent with joint principles; service doctrine introduces key joint topics; and linkages to joint documents and processes are present. Tone could range from service-centric to joint. Criteria for a joint tone are the following: describing the joint context and the joint force commander's role; portraying the service as a unique, rather than superior, contributor; using collaborative language; and portraying other services as contributors to a joint team.

Interviews with doctrine developers revealed that there are several reasons why Air Force and joint doctrine might be divergent. The Air Force's preference for brevity and use of documents other than operational doctrine, such as Air Force instructions, explain some *gaps* (defined as lack of discussion of key joint concepts) in Air Force doctrine. In other cases, joint doctrine might have been recently published, and Air Force doctrine might not have been updated yet to reflect any changes. Finally, the Air Force sometimes presents new or different concepts in its service doctrine as part of its effort to advocate for changes in joint doctrine. Although these considerations may justify divergence, this report describes all differences, regardless of the reasons behind them.

Assessments of Doctrinal Alignment

We assessed doctrinal alignment by comparing unclassified Air Force and joint operational doctrine documents using the above framework. The following briefly summarizes the results of our analysis. Five key conclusions arose from our assessment of substantive alignment. First, the extent of substantive alignment with joint doctrine varies. Some, but not all, Air Force annexes present principles and processes that are consistent with joint doctrine and use joint terms. Second, although the Air Force utilizes the joint publication numbering scheme, the contents of documents with the same number do not always align. Third, substantive gaps between Air Force and joint doctrine are a common reason for divergence. The Air Force strives for brevity and seeks to avoid repetition, but this sometimes results in doctrine documents that fail to introduce key joint concepts or to make linkages to other essential information. Moreover, Air Force doctrine does not consistently place topics within the broader context of a joint operation. Given the importance of planning in joint leadership positions, gaps in Air Force doctrine on planning processes and considerations are particularly notable.

Fourth, the Air Force employs some alternate constructs from those used in joint doctrine, sometimes without noting the differences. Consequently, service members may not always know which constructs are service-specific or how alternative constructs relate to those presented within joint doctrine. In some instances, Air Force doctrine also uses different terms than joint doctrine.

Finally, differences in the internal organization of the doctrine documents make it difficult to identify related information. Joint doctrine follows a standardized, nested framework for all its documents, helping readers to easily identify subjects and supporting materials. Comparatively, the Air Force's topic catalog structure makes it more difficult to understand how pieces of doctrine relate to each other or to joint doctrine.

Our analysis of Air Force doctrine's tone led to five major findings. First, some sections of Air Force doctrine displayed a highly joint tone, effectively incorporating joint context and describing the service as a unique, but not superior, contributor to joint teams. Annexes on special operations, foreign internal defense, and air mobility serve as excellent examples of a joint tone within Air Force doctrine. Second, some sections of Air Force doctrine were more insular, focusing primarily on service-specific issues with little discussion of the role the Air Force would play as part of a joint team. The electronic warfare and cyber operations annexes, for example, focus on how these tools support air operations, rather than joint operations. Third, some historical vignettes and senior leaders' quotes contribute to a service-centric tone. Fourth, some documents describe airmen as superior to their service counterparts as they advocate for Air Force positions. For example, Air Force doctrine advocates for airmen to command cyber and space operations because, the doctrine asserts, they take a broader view of war. Finally, the airmen's perspective sections often do not clearly state how and why the Air Force outlook differs so much from other services as to warrant the section's inclusion. Although declaring the value and unique contributions of the Air Force within service doctrine is understandable, indicating service superiority and focusing on service distinctions could also encourage the development of service-centric mindsets among airmen.

Doctrine in the Air Force

The doctrine comparisons revealed that there are certainly opportunities to strengthen the alignment of Air Force and joint doctrine. However, our analysis of the role of doctrine in the Air Force suggests that doctrinal revisions may only have a limited effect. A review of the history of doctrine within the Air Force showed that operational doctrine has not always been a service priority. Prior to 1993, responsibility for doctrine shifted frequently, and operational doctrine was rarely updated. Since 1993, operational doctrine development has been centrally managed by a doctrine center, now called the Curtis E. LeMay Center for Doctrine Development and Education.

Despite these organizational changes, there are indications that operational doctrine is still not an Air Force priority. Staffing levels have fallen since the LeMay Center's founding, and it does not have the resources to ensure that Air Force tactical doctrine is consistent with operational doctrine. Moreover, because of staffing cuts, Air Combat Command is no longer participating in the formal doctrine development process. Other Air Force major commands have also reduced the number of personnel devoted to doctrine development. Our review of the Air Force's primary professional journal indicates that doctrine is not a frequent topic of debate. On net, because doctrine is not an Air Force priority, doctrinal changes can only reinforce, not lead, the effort to enhance joint warfighting proficiency.

Experience with Joint Doctrine and Constructs

We also conducted exploratory analysis on how airmen encounter joint doctrine and constructs throughout their careers. Our analysis included interviews with 24 airmen currently or previously in joint assignments from across the Air Force, ranging in rank from major to major general. These airmen reported that they were exposed to joint doctrine too late and with inadequate opportunities to use joint doctrine in practice prior to joint assignments. This situation arises from several factors: Although very familiar with tactics, techniques and procedures, many airmen do not use operational doctrinal concepts until later in their careers; airmen generally have less experience with operational planning than their service counterparts; later exposure to joint settings may affect familiarity with joint constructs; and doctrine education and application are not always linked. These findings are not definitive, but they do suggest that joint proficiency may require providing airmen with more opportunities to use joint doctrine and constructs in practice throughout their careers.

Recommendations

Our analysis suggests that there are opportunities to increase the alignment of Air Force and joint doctrine. For example, the Air Force could more consistently introduce and summarize key joint concepts, use graphics to show relationships between Air Force and joint constructs, and provide more information on planning processes and considerations. Before the Air Force adopts these recommendations, it should consider whether some divergences should be preserved because of the unique characteristics of airpower or whether changes to joint doctrine would be more appropriate. Air Force leaders will also need to consider potential trade-offs with other Air Force priorities, such as adopting consistent terminology across the air, space, and cyber domains.

Although revisions to doctrine could be a first step, our analysis suggests that airmen may need additional opportunities to use joint doctrine and constructs in practice. The Air Force could provide these opportunities by introducing more joint terminology into tactical publications and activities and looking for opportunities to replace Air Force–specific planning terms and processes with their joint equivalents. Linking formal exposure to joint doctrine with joint experience may also help airmen internalize the joint language and mindset.

Acknowledgments

We would like to thank our sponsor, Maj Gen Brian M. Killough; our study action officer, Lt Col Edmund Loughran; Col Bryan Cannady; and the Chief of Staff of the Air Force focus area staff for supporting this project and providing input. We also appreciate those who took the time to meet with us to discuss Air Force and joint doctrine: staff at the Curtis E. LeMay Center for Doctrine and Education; members of the Joint Force Development Directorate, Joint Staff; Air Force officers and civilians at several combatant and Air Force major commands; and faculty and students at Air University. RAND colleagues Timothy Conley, Bill Marcellino, Karl Mueller, and Paula Thornhill also provided helpful suggestions. In addition, Daniel Ginsberg, Austin Long, and Karen Wilhelm provided thorough and thoughtful reviews of an earlier draft. We thank Mark Hvizda for assembling this document.

Abbreviations

ACC	Air Combat Command
ACSC	Air Command and Staff College
ADP	Army Doctrine Publication
AETC	Air Education and Training Command
AFI	Air Force Instruction
AFSOC	Air Force Special Operations Command
AFSOF	Air Force special operations forces
AFTTP	Air Force tactics, techniques, and procedures
AI	air interdiction
AOC	air operations center
ASPJ	*Air & Space Power Journal*
ATC	air tasking cycle
ATO	air tasking order
C2	command and control
CCDE	centralized control and decentralized execution
CCT	component-critical target
CJCS	Chairman of the Joint Chiefs of Staff
CJCSI	Chairman of the Joint Chiefs of Staff Instruction
COMAFFOR	commander, Air Force forces
C-RAM	counter-rocket, artillery, and mortar
CSAF	Chief of Staff of the Air Force
DoD	U.S. Department of Defense
DTM	doctrine topic module
EW	electronic warfare
FID	foreign internal defense
FSCL	fire support coordination line
FSCM	fire support coordination measure
GIISR	global integrated intelligence, surveillance, and reconnaissance
HD	homeland defense

IAMD	integrated air and missile defense
ISR	intelligence, surveillance, and reconnaissance
J-7	Joint Force Development Directorate of the Joint Staff
JAWS	Joint Advanced Warfighting School
JDDC	joint doctrine development community
JFACC	joint forces air component commander
JFC	joint force commander
JIPOE	joint intelligence preparation of the operating environment
JOPP	joint operation planning process
JP	joint publication
JPME	joint professional military education
JTC	joint targeting cycle
JTF	joint task force
LOE	line of effort
MAJCOM	major command
MD	missile defense
NOTAM	notice to airmen
NTC	National Training Center
OCA	offensive counterair
ROMO	range of military operations
SAASS	School for Advanced Air and Space Studies
SOF	special operations forces
TCT	time-critical target
TST	time-sensitive target
TTP	tactics, techniques, and procedures
USSOCOM	U.S. Special Operations Command
WMD	weapons of mass destruction

1. Introduction

Potential U.S. adversaries are fielding a range of advanced military capabilities, such as long-range precision strike, integrated air defenses, and offensive cyber capabilities, that are eroding long-standing U.S. military advantages. With air, land, sea, space, and cyberspace domains becoming increasingly contested, the United States has begun developing new operational concepts to regain its advantage. In particular, both joint and Air Force concepts envision gaining windows of superiority by integrating U.S. military capabilities more rapidly across multiple domains.[1] Joint operations, in this vision of the future, will require much deeper integration among the services than in the past. Implementing these concepts would also require the ability to integrate capabilities across domains at lower echelons of command.

In light of this growing demand for higher levels of joint integration, Chief of Staff of the Air Force (CSAF) Gen David Goldfein has launched a broad initiative to prepare airmen to integrate into, influence, and lead joint teams. Secretary of the Air Force Heather Wilson reinforced the importance of this initiative, stating that "we need leaders who can thrive in joint teams."[2] As a first step, Goldfein asked 9th Air Force to develop the capability to become a deployable joint task force (JTF) headquarters.[3] The CSAF also called on the Air Force to "reinvigorate our development to purposefully and systematically gain proficiency in joint warfare earlier in the careers of Airmen."[4]

Unfortunately, there is little quantitative information about the status of joint proficiency in the Air Force. Therefore, the CSAF's initiative is motivated by an assumption that joint proficiency, wherever the level currently stands, could be improved. The Air Force is considering many possible ways to improve airmen's joint warfighting proficiency, including changes to Air Force doctrine.

---

[1] U.S. Air Force, *Air Force Future Operating Concept: A View of the Air Force in 2035*, Washington, D.C., 2015, pp. 8–9; U.S. Department of Defense (DoD), *Joint Operational Access Concept (JOAC)*, Washington, D.C., January 17, 2012, pp. 17, 21; Michael E. Hutchens et al., "Joint Concept for Access and Maneuver in the Global Commons: A New Joint Operational Concept," *Joint Forces Quarterly*, Vol. 84, First Quarter 2017.

[2] Heather Wilson, "State of the Air Force," presentation at 2017 Air Force Association Air, Space & Cyber Conference, National Harbor, Md., September 18, 2017.

[3] Amanda Dick, "Commander Sets Priorities, Way Ahead for 9th AF," *News: 9th Air Force*, February 2, 2017.

[4] Dave Goldfein, *CSAF Focus Area: Strengthening Joint Leaders and Teams*, Washington, D.C.: U.S. Air Force, 2016. For further discussions of Air Force officer preparation for joint assignments see, Caitlin Lee et al., *Rare Birds: Understanding and Addressing Air Force Underrepresentation in Senior Joint Positions in the Post–Goldwater Nichols Era*, Santa Monica, Calif.: RAND Corporation, RR-2089-AF, 2017, pp. 16–17; Michael Spirtas, Thomas-Durrell Young, and S. Rebecca Zimmerman, *What It Takes: Air Force Command of Joint Operations*, Santa Monica, Calif.: RAND Corporation, MG-777-AF, 2009, pp. 10–11.

The Air Force asked RAND Corporation researchers to look specifically at differences between Air Force and joint doctrine and to identify revisions to Air Force doctrine that could prepare airmen to operate in the more deeply integrated joint settings of the future. *Joint doctrine* is defined as the "fundamental principles that guide the employment of United States military forces in coordinated action toward a common objective and may include terms, tactics, techniques, and procedures."[5] But joint doctrine is more than just the operational approach to warfare that prevails when the services fight together. Joint doctrine also establishes the language of the joint community: the terms, relationships, organizations, processes, and principles that the services have all agreed to use in joint settings. These elements of joint doctrine, which we generically refer to as *joint constructs*, can also be used by single services and at other levels of war.[6] Joint doctrine also defines the *joint mindset*, which is the idea that the services integrate, not just coordinate or deconflict, when they fight as part of a joint team.

Learning joint constructs and adopting an integrated joint mindset are a necessary, but not sufficient, condition for integrating into, influencing, and leading future joint teams. Much like becoming a highly skilled pilot, developing high levels of joint proficiency requires more than knowing the right language; it requires extensive practice and operational experience.[7] Still, habitual use of the joint language and early adoption of the joint mindset should lay the groundwork for becoming a skilled joint warfighter.

There are two benefits to looking at the alignment of joint and Air Force doctrine. First, unless doctrine is undergoing significant revision, it is generally considered a good snapshot of the extant practices and culture of a service. A comparison of Air Force and joint doctrine therefore offers one way to assess the Air Force's current use and acceptance of joint doctrine and constructs. Chapter 2 develops a framework for comparing Air Force and joint doctrine and presents the results of detailed document comparisons. The results suggest that the adoption of the joint language and mindset in the Air Force is uneven.

Second, in theory, revisions to Air Force doctrine to increase the alignment with joint doctrine could be part of a broader initiative to promote joint warfighting proficiency. That does not mean that Air Force doctrine needs to be identical to or perfectly consistent with joint doctrine on every point. The Air Force could, for example, choose to maintain certain doctrinal differences because of the unique characteristics of airpower. However, minimizing and explaining inconsistencies between Air Force and joint doctrine could potentially help airmen to

---

[5] DoD, *DoD Dictionary of Military and Associated Terms*, Washington, D.C., March 2017, p. 125.

[6] Chairman of the Joint Chiefs of Staff Instruction (CJCSI) 5120.02D, *Joint Doctrine Development System*, Washington, D.C.: Joint Chiefs of Staff, January 5, 2015, pp. A-2. Our review of doctrine found that it can also include intellectual frameworks, philosophies, principles, and other elements.

[7] On higher levels of proficiency being developed through a combination of education, training, and experience, see Air Force Instruction (AFI) 10-2801, *Force Development Concepts*, Washington, D.C.: Headquarters, U.S. Air Force, October 23, 2014; Annex 1-1, *Force Development*, Washington, D.C.: Headquarters, U.S. Air Force, April 17, 2017.

speak the same language as their joint counterparts and gain familiarity with the processes and principles they will encounter in joint settings.

However, determining whether revisions to Air Force doctrine would affect joint proficiency in practice requires additional analysis. Therefore, Chapter 3 considers how airmen currently encounter service and joint doctrine to answer two questions. First, to what extent could greater alignment of Air Force and joint doctrine help airmen gain greater joint proficiency? Second, would changes to the way Air Force officers learn or use doctrine help to promote joint proficiency? To answer these questions, we collected data on the resources devoted to doctrine development and the discourse on doctrine within a key Air Force professional journal. Chapter 3 also uses several exploratory approaches to identify other conditions that affect airmen's familiarity with joint constructs. This analysis included interviews with those who teach doctrine as well as airmen with current or recent joint experience. We focused on active-duty Air Force officers, conducted a limited number of interviews, and used readily available data, so the analysis is not exhaustive. Still, the findings highlight many key issues that the Air Force will need to consider as it takes steps to increase joint proficiency.

Chapter 4 provides recommendations for improving the joint warfighting proficiency of Air Force officers, with an emphasis on the potential contributions of Air Force doctrine. Our in-depth analysis of Air Force doctrine allowed us to make detailed recommendations about possible revisions to Air Force doctrine and the doctrine development process. Drawing on our exploratory analysis in Chapter 3, we also identify general recommendations for force development and identify trade-offs policymakers should consider as they take steps to promote joint proficiency.

Before presenting the analysis and recommendations, the remainder of this introduction provides background on Air Force and joint doctrine, as well as the processes by which doctrine is written.

Operational Doctrine and Doctrine Development

This study focuses on Air Force and joint operational doctrine. What follows is a brief discussion of what operational doctrine is and how both Air Force and joint doctrine are developed. Operational doctrine consists of principles and practices for warfare at the operational level, traditionally drawn from combat experience, analysis, wargames, and exercises. Air Force and joint doctrine generally focus on validated best practices rather than new ideas for how operations could be conducted in the future, known as concepts.[8] Tactics, techniques, and

---

[8] CJCSI 5120.02D, 2015. Air Force doctrine states that

> operational doctrine guides the proper organization and employment of air, space, and cyberspace forces in the context of distinct objectives, force capabilities, broad functional areas, and operational environments. Operational doctrine provides the focus for developing the missions and tasks to be executed through tactical doctrine. (LeMay Center, *Core Doctrine,* Vol. I, *Basic Doctrine*, Washington, D.C.: Headquarters, U.S. Air Force, February 27, 2015a, p. 19)

procedures (TTP) documents—sometimes referred to as tactical doctrine—cover more-technical issues and are not discussed in detail in this report.[9]

Both Air Force and joint doctrine documents describe doctrine as authoritative and requiring judgment in application. However, there are some subtle differences in the language they use about the applicability of doctrine. Joint Publication (JP) 1, *Doctrine of the Armed Forces of the United States*, states that joint doctrine should generally be followed, "except when, in the judgment of the commander, exceptional circumstances dictate otherwise."[10] Air Force doctrine, on the other hand, puts less emphasis on this final point. Air Force documents state that doctrine is "authoritative, not directive and requires judgment in application" and describe it as an "informed starting point."[11]

There are many similarities in the processes for developing joint and Air Force doctrine. In general, both processes are centrally managed by doctrine developers, but the inputs for the substance of doctrine come from the larger communities.

Joint Doctrine Development

Joint doctrine documents, JPs, are used by many audiences but are written especially for those involved in the operational level of war, such as combatant commanders, joint task force commanders, and their staffs. A key purpose of doctrine is to develop a common language and framework for those fighting together. As a result, service doctrine is supposed to be consistent with joint doctrine—and joint doctrine "takes precedence over" service doctrine in joint settings.[12]

The Joint Force Development Directorate of the Joint Staff (J-7) manages joint doctrine development for the Chairman of the Joint Chiefs of Staff (CJCS). The joint doctrine development community (JDDC) includes such organizations as the combatant commands, the services, and directorates of the Joint Staff. The J-7 does not write most JPs. Instead, a lead agent from within the

---

[9] Air Force doctrine states that

> tactical doctrine describes the proper employment of specific Air Force assets, individually or in concert with other assets, to accomplish detailed objectives. Tactical doctrine considers particular objectives (stopping the advance of an armored column) and conditions (threats, weather, and terrain) and describes how Air Force assets are employed to accomplish the tactical objective (B-1 bombers dropping anti-armor cluster munitions). (LeMay Center, 2015a, p. 19)

Air Force TTP (AFTTP) documents are coordinated through other organizations, such as the 561st Joint Tactics Squadron and the 423 Mobility Training Squadron. Multiservice TTP documents are coordinated through the Air, Land, Sea Applications Center (ALSA); U.S. Air Force Expeditionary Center, "423d Mobility Training Squadron," August 20, 2015; Nellis Air Force Base, "561st Joint Tactics Squadron," fact sheet, March 24, 2016.

[10] JP 1, *Doctrine of the Armed Forces of the United States*, Washington, D.C.: Joint Chiefs of Staff, March 25, 2013, p. ii. This is also stated in policy; CJCSI 5120.02D, 2015, p. A-2.

[11] LeMay Center, 2015a, pp. 2, 10; AFI 10-1301, *Air Force Doctrine Development*, Washington, D.C.: Headquarters, U.S. Air Force, June 14, 2013, incorporating Change I, April 23, 2014; AFI 10-13, *Air Force Doctrine*, Washington, D.C.: Headquarters, U.S. Air Force, August 25, 2008.

[12] CJCSI 5120.02D, 2015.

JDDC is assigned to develop each JP. The J-7 manages the process for requesting and receiving inputs from and approval of the JDDC for each new document or revision.[13]

Air Force Doctrine Development

Air Force operational doctrine is made up of three volumes of *Core Doctrine* and 29 annexes.[14] Although these documents are also used by other audiences, they are written primarily to inform those who lead Air Force forces in warfighting at the operational level; a commander, Air Force forces (COMAFFOR); and his or her staff. Air Force doctrine is web-based, and each volume or annex consists of a series of short discussions of specific doctrine topics called *doctrine topic modules* (DTMs).[15]

The Curtis E. LeMay Center for Doctrine and Education, hereafter referred to as the LeMay Center, is the CSAF's executive agent for doctrine and is under the administrative control of Air University. The LeMay Center's doctrine development directorate manages Air Force doctrine development and provides service inputs to joint and multinational doctrine. The directorate's staff are experts in doctrine development, but they are not necessarily subject-matter experts in the doctrine they oversee. Instead, the substance of doctrine is drawn from operators' inputs within the Air Force's major commands (MAJCOMs). When new documents are written or doctrine is revised, the LeMay Center requests input from the MAJCOMs. A doctrine point of contact at each MAJCOM requests input from subject-matter experts within their command and compiles responses for input to the LeMay Center. The LeMay Center staff integrates those inputs and publishes the updated doctrine. During this process, the LeMay Center reviews Air Force doctrine for consistency with joint doctrine.[16] However, as described in more detail in Chapter 2, there are a number of reasons why Air Force doctrine sometimes diverges from joint doctrine.

The LeMay Center is also the Air Force's representative for joint doctrine issues. In this role, it acts as the lead agent for five doctrine documents and coordinates Air Force inputs to other JPs.[17]

---

[13] CJCSM 5120.01A, *Joint Doctrine Development Process*, Washington, D.C.: Joint Chiefs of Staff, December 29, 2014.

[14] Vols. I and II of *Core Doctrine* are basic doctrine, and Vol. III is a summary of key issues in the annexes; LeMay Center, 2015a; LeMay Center, *Core Doctrine,* Vol. II, *Leadership,* Washington, D.C.: Headquarters, U.S. Air Force, August 8, 2015b; LeMay Center, *Core Doctrine,* Vol. III, *Command*, Washington, D.C.: Headquarters, U.S. Air Force, November 22, 2016.

[15] Curtis E. LeMay Center for Doctrine Development and Education, homepage, undated.

[16] AFI 10-1301, 2013, incorporating Change I, April 23, 2014.

[17] These documents are JP 3-03, *Joint Interdiction*, Washington, D.C.: Joint Chiefs of Staff, September 9, 2016; JP 3-30, *Command and Control of Joint Air Operations*, Washington, D.C.: Joint Chiefs of Staff, February 10, 2014; JP 3-52, *Joint Airspace Control*, Washington, D.C.: Joint Chiefs of Staff, November 13, 2014; JP 3-59, *Meteorological and Oceanographic Operations*, Washington, D.C.: Joint Chiefs of Staff, December 7, 2012; JP 3-60, *Joint Targeting*, Washington, D.C.: Joint Chiefs of Staff, January 31, 2013.

2. Joint Language and Mindset in Air Force Doctrine

Existing Air Force doctrine is a useful indicator of the extent to which joint terms, processes, and other constructs have been adopted in practice. Air Force doctrine development is centrally managed by the LeMay Center, but the content is drafted and revised by expert practitioners within Air Force MAJCOMs.[1] Air Force doctrine is not directive, so it may not reflect all variations in current practice.[2] Still, as the "officially sanctioned approach" to military operations, Air Force doctrine provides an imperfect, but important, indicator of current practice.[3] Studies of military doctrine have also found that an organization's culture deeply affects its doctrine.[4] As a result, the way that existing doctrine discusses other services and joint operations is also an indicator of the broader beliefs about "jointness" within the Air Force.

This chapter develops frameworks for assessing the substantive alignment between Air Force doctrine and joint doctrine, as well as the tone of Air Force doctrine. Detailed comparisons of Air Force and joint doctrine documents showed that Air Force doctrine documents vary in the extent to which they use joint constructs or link Air Force operations to the broader joint context. Moreover, the tone of Air Force doctrine also varies: Some documents characterize the Air Force as part of a joint team, while others had a service-centric orientation. To the extent that Air Force doctrine is a reflection of current practice, the results suggest that joint language and a joint mindset have only been partially adopted.

Doctrine Comparison Methodology

We conducted a detailed qualitative comparison of Air Force and joint doctrine documents using standardized frameworks described below. Although Air Force doctrine annexes follow the joint numbering scheme, documents with the same number do not always discuss the same subjects. Therefore, we identified and then compared documents that covered the same topics (see Table 2.1). For example, we compared Annex 1-1, *Force Development*, with a section of JP 1, *Doctrine of the Armed Forces of the United States*, because they both discuss force

---

[1] AFI 10-1301, 2013, incorporating Change I, April 23, 2014; LeMay Center, in-person, phone, and email discussions with RAND project team, Maxwell Air Force Base, Ala., July–November 2017c.

[2] LeMay Center, 2015a, p. 8.

[3] Paul Johnston, "Doctrine Is Not Enough: The Effect of Doctrine on the Behavior of Armies," *Parameters*, Autumn 2000.

[4] *Culture* is the "shared beliefs about the organization and its mission." Austin Long, *The Soul of Armies: Counterinsurgency Doctrine and Military Culture in the US and UK*, Ithaca, N.Y.: Cornell University Press, 2016, pp. 15, 19; Dima Adamsky, *The Culture of Military Innovation: The Impact of Cultural Factors on the Revolution in Military Affairs in Russia, the US, and Israel*, Stanford, Calif.: Stanford University Press, 2010.

development.[5] The report occasionally references other types of publications, such as Air Force tactics, techniques, and procedures (AFTTP) documents and AFIs. However, we did not systematically review this vast set of documents.

Table 2.1. Joint and Air Force Doctrine Publication Comparisons

Topic	Air Force Document(s)	Joint Publication(s)
Basic doctrine	*Core Doctrine*, Vol. I	JP 1, JP 3-0
Leadership	*Core Doctrine*, Vol. II	None
Legal support to operations	Annex 1-04	JP 1-04
Force development	Annex 1-1	JP 1
Intelligence	Annex 2-0	JP 2-0, JP 2-01, JP 2-01.3, JP 2-03
Operations and planning	Annex 3-0	JP 3-0, JP 5-0
Counterair operations	Annex 3-01	JP 3-01
Counterland operations	Annex 3-03	JP 3-03, JP 3-09, JP 3-09.3
Countersea operations	Annex 3-04	JP 3-02, JP 3-04, JP 3-32
Special operations	Annex 3-05	JP 3-05
Irregular warfare	Annex 3-2	JP 3-05.1, JP 3-07, JP 3-20, JP 3-22, JP 3-24, JP 3-26
Protection	Annex 3-10	JP 3-0, JP 3-10, JP 4-02
Cyberspace operations	Annex 3-12	JP 3-12(R), JP 6-0
Information operations	Annex 3-13	JP 3-13, JP 3-13.2, JP 3-13.3, JP 3-13.4
Space operations	Annex 3-14	JP 3-14
Air mobility	Annex 3-17	JP 3-17
Foreign internal defense	Annex 3-22	JP 3-22, JP 3-24
Homeland operations	Annex 3-27	JP 3-27, JP 3-28
C2	Annex 3-30	JP 1, JP 3-30
Engineering operations	Annex 3-34	JP 3-34
Countering WMD	Annex 3-40	JP 3-40
Personnel recovery	Annex 3-50	JP 3-50
Electronic warfare	Annex 3-51	JP 3-13.1
Airspace control	Annex 3-52	JP 3-52
Weather operations	Annex 3-59	JP 3-59
Targeting	Annex 3-60	JP 3-60
Public affairs	Annex 3-61	JP 3-61
Strategic attack	Annex 3-70	JP 3-0, DoDD 5100.01
Nuclear operations	Annex 3-72	None
Logistics	Annex 4-0, Annex 4-02	JP 4-0, JP 4-01, JP 4-01.2, JP 4-01.5, JP 4-01.6, JP 4-02, JP 4-03, JP 4-05, JP 4-06, JP 4-08, JP 4-09, JP 4-10

NOTE: *Core Doctrine*, Vol. III (LeMay, 2016), was not included in our review, because it contains summaries of the information found in the annexes. C2 = command and control; DoDD = Department of Defense Directive; WMD = weapons of mass destruction.

[5] Annex 1-1, 2017; JP 1, 2013.

Framework for Document Comparisons

For the detailed document comparisons, we developed a standardized framework for comparing Air Force and joint doctrine. The elements of this framework draw on the main goals of doctrine outlined in joint and service doctrine, as well as writing by outside analysts.[6]

First, we looked at the organization of doctrine within a given issue area (e.g., space operations, cyber operations), both at the document level and within the documents. Internally, we asked whether there were substantial differences in the way the documents are organized that reflected important differences in the logic behind Air Force and joint doctrine. We also considered whether the organization of Air Force doctrine reinforced connections with joint doctrine.

Second, we looked at how each document characterized the operating environment. In other words, whether the documents had different visions about the international system, the threat environment, enemy capability, or the missions the U.S. military would be called on to accomplish.

Third, we considered the operational principles that were explicitly or implicitly motivating doctrine. We asked whether the doctrine has consistent views about ends, ways, and means, as well as assumptions about cause and effect. In particular, we looked at whether Air Force and joint doctrine had consistent views about the role of airpower in joint operations.

Fourth, in addition to having a common picture of how military force should be used, a joint team needs to have common processes to achieve its goals. Therefore, we looked at whether the processes presented in Air Force doctrine were consistent with those outlined in joint doctrine and whether key joint processes were introduced.

Fifth, we asked whether Air Force doctrine identified the same roles and responsibilities as joint doctrine.

Finally, we looked at whether the documents used the same terminology to allow airmen to speak the language of a joint environment. We examined whether Air Force doctrine used service-specific terms when joint terms existed and whether they used joint terms in the same manner as joint documents.

Assessing Alignment of Substance and Tone

After reviewing doctrine documents using the framework, we assessed the overall alignment of Air Force and joint doctrine using two scales: one that assessed the substance of doctrine (see Figure 2.1) and a second that looked at the document's tone (see Figure 2.2). The ranges of these

---

[6] LeMay Center, 2015a, p. 10; Headquarters, Department of the Army, *Doctrine Primer*, Washington, D.C., ADP 1-01, September 2014b; Barry R. Posen, *The Sources of Military Doctrine: France, Britain, and Germany Between the World Wars*, Ithaca, N.Y.: Cornell University Press, 1984, p. 13; David E. Johnson, *Learning Large Lessons: The Evolving Roles of Ground Power and Air Power in the Post–Cold War Era*, Santa Monica, Calif.: RAND Corporation, MG-405-1-AF, 2007, p. xix.

scales do not represent the full theoretical range of differences that any two pieces of doctrine could have: from fundamentally opposing premises about warfare to identical documents. U.S. joint and service doctrine share many of the same basic assumptions. At the same time, these documents serve different purposes, so they should not simply repeat the same information. As a result, these scales represent a way to compare Air Force doctrine relative with one another somewhere in the middle of the larger theoretical range. The two scales could be used to assess the alignment between any U.S. service's doctrine and joint doctrine.

Substance

The first scale assesses the substantive alignment of the content of service and joint doctrine through the application of five measures: (1) consistency of key principles and processes, (2) coverage of key topics, (3) connections to joint documents and processes, (4) application of joint processes, and (5) use of joint terminology. The factors are listed in order of importance, with the most important factor listed first. Thus, consistency of key principles is the most important measure for alignment. Inconsistencies between service and joint principles, especially if the service doctrine does not explain the inconsistencies, counter the goal of providing service members with the necessary tools to succeed in joint settings. When service doctrine has gaps (meaning key joint constructs are not introduced), service members could be unprepared to operate in a joint environment. Service doctrine does not need to repeat all the content found in joint doctrine, but it should at least briefly introduce key joint concepts and provide links to relevant joint documents. Use of joint terms, although less critical than operational concepts and processes, could also help service members speak the same language as their joint counterparts.

Tone

We also assessed each document to determine whether its overall tone reflected a joint mindset. In 1986, the Goldwater-Nichols Act mandated structural reforms to DoD in an attempt to reduce interservice rivalry and increase the effectiveness of joint operations. In 1993, then-CJCS Colin Powell described *jointness* as follows:

> Our soldiers know that they are the best on the battlefield; our sailors know that they are the best at sea; our airmen know that they are the finest in the skies; our Marines know that no one better ever hit the beach. But every one of these men and women also knows that they play on a team. They are of the team and for the team; "one for all and all for one," as Alexandre Dumas put it in *The Three Musketeers*. We train as a team, fight as a team, and win as a team.[7]

[7] Colin L. Powell, "A Word from the Chairman," *Joint Force Quarterly*, Vol. 1, No. 1, 1993.

Figure 2.1. Substantive Alignment Scale

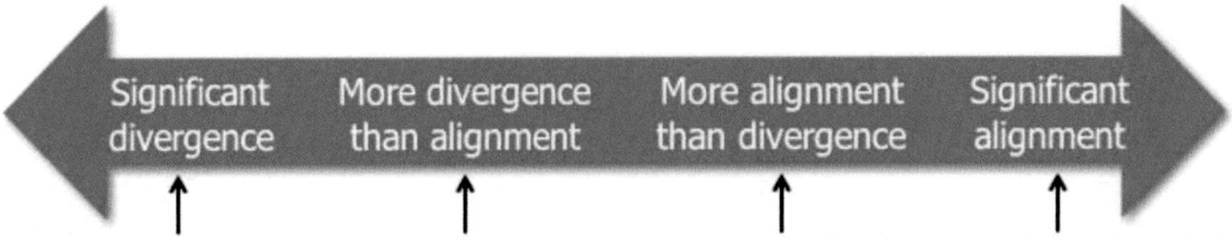

Significant divergence

- Key service principles **conflict with** joint principles
- Service doctrine overlooks many key topics and/or covers them in insufficient detail
- Links to joint documents and processes are absent
- Service applies conflicting processes
- Service uses mainly different terms

More divergence than alignment

- Key service principles **differ notably from** joint principles
- Service doctrine overlooks several key joint topics and/or covers them in insufficient detail
- Some links to joint documents and processes are present
- Service applies different processes
- Service uses many different terms

More alignment than divergence

- Key service principles **differ somewhat from** joint principles
- Service doctrine overlooks a few key topics and/or covers them with insufficient detail
- Many links to joint documents and processes are present
- Service applies similar processes
- Service uses a few different terms

Significant alignment

- Key service principles **are consistent with** joint principles
- Air Force covers most key topics with sufficient detail
- Links to joint documents are present consistently
- Service applies joint processes
- Service doctrine mainly uses joint terms

The second scale (Figure 2.2) builds on this basic concept of jointness and assesses whether the tone of service doctrine is joint or service-centric using four measures: (1) inclusion of joint context and reference to the joint force commander, (2) portrayal of the service's contributions to joint operations, (3) depiction of other services, and (4) whether the overall tone was collaborative. As with the previous scale, the criteria for the tone scale are listed in order of importance, with the presence of joint context and references to the joint force commander as the most important of the four measures. The fundamental tenets of a joint force are that the joint force commander (JFC) is the supported commander and that the services operate as a joint team.

Figure 2.2. Tone Scale

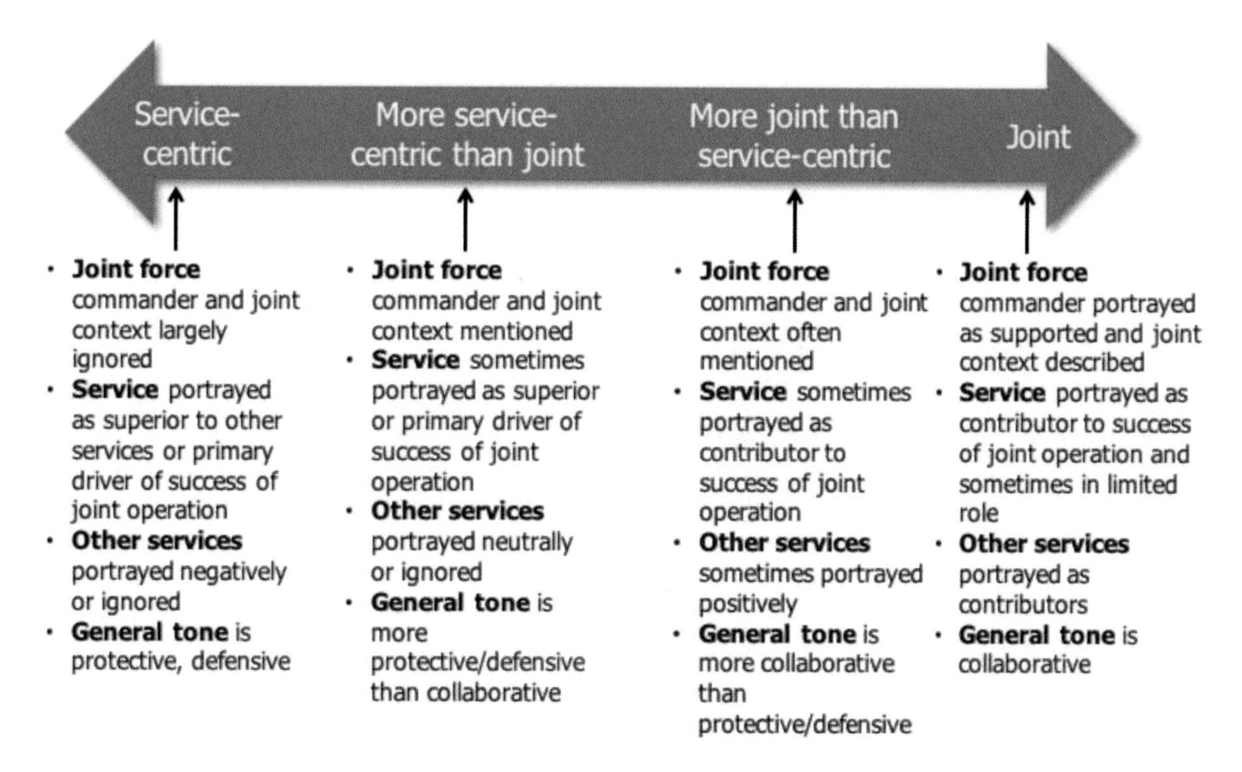

Air Force doctrine may wish to include the joint perspective for two reasons. First, including references to a joint context gives airmen practical information about operating in a joint setting. Second, by referencing key joint constructs throughout service doctrine, the service implicitly endorses them, making these constructs more likely to be an accepted part of the service's vocabulary and mindset. The context in which the Air Force portrays itself compared with other services, the second measure of tone, lays the foundation for how airmen see themselves while operating in joint settings. Doctrine with a joint, rather than service-centric, tone reinforces the notion that the services make unique, but not superior, contributions as part of a mutually reinforcing joint team. The additional two measures identify ways in which Air Force doctrine can subtly encourage a service-centric or joint mindset. Like the alignment scale, our research remained agnostic on the reasons for including language with differing tones. If there was a mix of tones in the same document, we used judgment to determine an overall tone assessment based on the order of the criteria listed above.

Sources of Divergence

Decisions about whether any given piece of Air Force doctrine should be brought into alignment with joint doctrine should take a range of factors into account. However, as a first step toward informing decisions about potential changes to its doctrine, the Air Force needs to understand where differences exist and how substantial they are. Therefore, our research assesses the alignment of Air Force doctrine with joint doctrine regardless of the reason for any misalignment.

However, we are aware that there are several considerations that explain some of the divergence we describe in this report. First, updates to the Air Force doctrine may not coincide with updates to joint doctrine—therefore, there could be differences because of timing.[8] Second, Air Force doctrine is intentionally written to be brief to encourage more airmen to read it. The LeMay Center actively seeks to reduce repetition with joint doctrine. Therefore, in general, Air Force annexes are not comprehensive, stand-alone documents. Rather, readers generally need to be familiar with joint doctrine to understand Air Force doctrine.[9] In contrast, interviews with members of the J-7 revealed that joint doctrine documents are written to be stand-alone books. Although it often references other documents, joint doctrine provides the reader with context to understand key issues related to the doctrine document.[10]

Third, the Air Force covers some topics that are discussed in joint doctrine in other types of publications. For example, as discussed below, the Air Force covers some topics related to countering WMD in AFTTP documents. Similarly, some of the topics covered in J-7 are covered in an AFI rather than operational doctrine.

Fourth, the Air Force sometimes retains divergent constructs in its doctrine because it does not believe that joint doctrine is appropriate for the service. This could be because, for example, the Air Force believes that the unique characteristics of airpower necessitate a different approach.

Finally, the Air Force incorporates new constructs within its doctrine that it wants as part of joint doctrine. Showing commitment to a construct through its inclusion in service doctrine is an informal norm of the joint doctrine development process.[11] Setting aside these reasons behind Air Force variations allows us to ensure that we identify all differences for the service's consideration.

Doctrine Comparison Findings

Our comparative analysis of Air Force and joint doctrine resulted in two broad findings. First, acceptance of joint constructs within the Air Force is incomplete, as judged by the substantive alignment of Air Force doctrine. Some Air Force doctrine documents are aligned with joint doctrine, but many are not well aligned and diverge to varying degrees. We identified a few key ways in which Air Force doctrine diverges from joint doctrine: Some key constructs

---

[8] We did, however, find cases where divergences persisted because Air Force annexes had not been updated on the Air Force's standard two-year revision cycle. See, for example, Annex 3-12, *Cyberspace Operations*, Washington, D.C.: Headquarters, U.S. Air Force, November 30, 2011.

[9] LeMay Center, 2017c.

[10] J-7, in-person and email discussions with RAND project team, Pentagon, Washington, D.C., July–October 2017.

[11] LeMay Center, 2017c; J-7, 2017.

(such as C2 philosophies, the framework for integrating air and missile defense, and the approach to operational design) differ; gaps in coverage of key topics (such as joint functions and joint planning) exist; and links to joint documents and processes are absent. In addition, overall differences in organization suggest that showing the links between Air Force and joint doctrine is not a central consideration.

Second, the tone of Air Force doctrine also varies. Some Air Force doctrine displays a highly joint tone, effectively incorporating joint context and describing the service as a unique, but not superior, contributor to joint teams. Other Air Force doctrine, however, reflects a service-centric mindset. This second finding is relevant because, even if Air Force doctrine universally used joint constructs, it could still impart a mindset that inhibits integration with other services. Because doctrine is historically an important reflection of service culture and practice, these findings suggest that acceptance of a joint mindset within the Air Force is uneven. The following sections describe each of these findings in detail, using select examples from our detailed document comparisons.

In this report, we devote more space to discussing where Air Force doctrine diverges from joint doctrine and takes on a service-centric tone than to discussing where it is aligned with joint doctrine and displays a joint tone. This is not meant to suggest that Air Force doctrine is on the whole more divergent and service-centric than it is aligned and joint; that is not the case, as subsequent figures will show. Instead, our intent is to point out those areas that could be improved in the interest of promoting joint proficiency.

This report does not assess the substance or tone of other services' doctrine or compare it with our findings from Air Force doctrine. Such analysis also would likely reveal some divergence in substance and presence of service-centric tone. The focus of this report, however, is to identify where the Air Force is today on use of joint language and a joint mindset, regardless of where other services stand.

Air Force Acceptance of Joint Constructs Is Incomplete

Overall, the first major finding is that current Air Force doctrine reflects an incomplete acceptance of joint doctrinal constructs. Figure 2.3 shows the summary of the alignment of Air Force doctrine with joint doctrine. The next three sections present examples of Air Force doctrine that fall into each category: significantly aligned with joint doctrine, more aligned with than divergent from joint doctrine, and more divergent from than aligned with joint doctrine.

Figure 2.3. Air Force Doctrine Alignment with Joint Doctrine

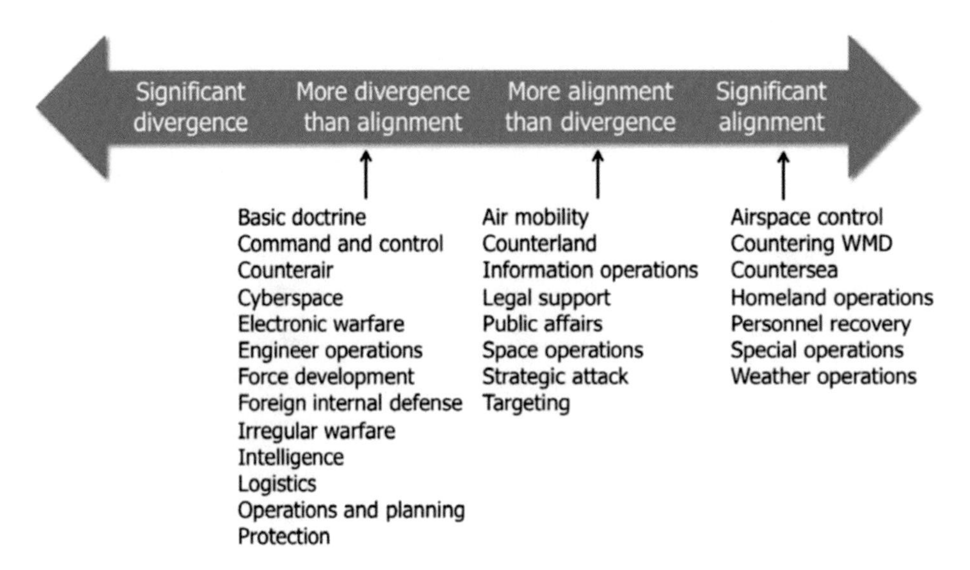

NOTE: We reviewed a total of 31 Air Force doctrine documents. The two logistics annexes, Annex 4-0 and Annex 4-02, are combined in this figure. *Core Doctrine,* Vol. II, *Leadership*, and Annex 3-72, *Nuclear Operations*, are not included in this figure because they have no joint equivalents to provide a basis for comparison (Annex 3-72, *Nuclear Options*, Washington, D.C.: Headquarters, U.S. Air Force, May 19, 2015; Annex 4-02, *Medical Operations*, Washington, D.C.: Headquarters, U.S. Air Force, September 29, 2015; Annex 4-0, *Combat Support*, Washington, D.C.: Headquarters, U.S. Air Force, December 21, 2015).

Some Air Force Doctrine Reflects Significant Alignment with Joint Doctrine

Seven Air Force annexes were significantly aligned with joint doctrine. Annex 3-05, *Special Operations*, is an example of a document that is well aligned. This annex describes the same principles and C2 relationships as joint doctrine and uses joint terms.[12] Also, a significant portion of the language used in Annex 3-05 to describe terms, concepts, and processes is nearly identical to the language in corresponding sections of JP 3-05.[13]

Annex 3-52, *Airspace Control*, is the Air Force counterpart to JP 3-52, *Joint Airspace Control*, and it is also closely aligned with joint doctrine.[14] The Air Force was the lead agent for the JP, which could explain the high level of alignment.[15] The key principles discussed in

---

[12] Annex 3-05, *Special Operations*, Washington, D.C.: Headquarters, U.S. Air Force, February 9, 2017; JP 3-05, *Special Operations*, Washington, D.C.: Joint Chiefs of Staff, July 16, 2014.

[13] Although Annex 3-05 uses very similar (in some cases, nearly identical) language to that of JP 3-05, it does not always cite JP 3-05 as the source. For examples of these similarities in language, see Annex 3-05, 2017, pp. 6–7, 8–10, 12–14, and 19–22.

[14] Annex 3-52, *Airspace Control*, Washington, D.C.: Headquarters, U.S. Air Force, July 21, 2014; JP 3-52, 2014.

[15] That being said, of the five JPs for which the Air Force is the lead agent, only two of the corresponding Air Force doctrine publications (*Airspace Control* and *Weather Operations*) display significant alignment with the JPs. Two others (*Targeting* and *Counterland Operations*) are more aligned than divergent, while the fifth, *C2*, is more divergent than aligned. Divergence in these three publications is, in part, because of gaps and poor linkages to joint

Annex 3-52 are consistent with joint principles for airspace control. There are consistent references to joint doctrine throughout the annex, and both doctrine documents describe complementary organizational structures. Both documents also specify which airspace authorities (e.g., aeronautical information publication, air tasking order [ATO], airspace control order) are applicable across the phases of an operation (from Phase 0 through Phase V).[16]

In another example of a document with significant alignment, Annex 3-27, *Homeland Operations*, deals with the same key topics as JP 3-27, *Homeland Defense*, and JP 3-28, *Defense Support of Civil Authorities*: national-level policy and guidance documents; statutes that affect military operations in the homeland; command relationships and authorities among military commands and components, as well as between military and civilian organizations; and rules of engagement and rules on the use of force in the homeland. The JPs are considerably longer, cover more topics, and include more detail on some topics than Annex 3-27. Although this annex covers fewer topics, no gaps on key issues exist.[17]

Some Air Force Doctrine Displays More Alignment with Than Divergence from Joint Doctrine

Eight annexes are more aligned with than divergent from joint doctrine: *Air Mobility Operations*, *Counterland Operations*, *Information Operations*, *Legal Support*, *Public Affairs Operations*, *Space Operations*, *Strategic Attack*, and *Targeting*. In the case of Annex 3-70, *Strategic Attack*, the Air Force view about strategic attack is generally consistent with other aspects of joint doctrine.[18] Annex 3-70 describes strategic attack as an approach to war that bypasses tactical and operational effects to directly achieve strategic effects and objectives. JP 3-0 also acknowledges that strategic attack "may achieve strategic objectives without necessarily having to achieve operational objectives as a precondition."[19] In that respect, Annex 3-70 is consistent with JP 3-0. However, there is divergence between the two documents on the issue of the Air Force's role in strategic attack. JP 3-0 specifically states that "all components of a joint

---

doctrine. However, differences in terminology and other constructs suggest that the Air Force might have made an intentional decision to maintain some service-specific constructs.

[16] Annex 3-52 (2014) also shows phases of operation as part of the diagram. Although not explicitly shown in the figure, JP 3-52 discusses phases of operation as well in a consistent manner; see Chapter III of JP 3-52, 2014.

[17] U.S. Northern Command was the lead agent for JP 3-27 and JP 3-28. Annex 3-27, *Homeland Operations*, Washington, D.C.: Headquarters, U.S. Air Force, April 28, 2016; JP 3-27, *Homeland Defense*, Washington, D.C.: Joint Chiefs of Staff, July 29, 2013; JP 3-28, *Defense Support of Civil Authorities*, Washington, D.C.: Joint Chiefs of Staff, July 31, 2013.

[18] There is no dedicated joint doctrine document that corresponds to Annex 3-70, *Strategic Attack*, Washington, D.C.: Headquarters, U.S. Air Force, May 25, 2017. However, JP 3-0 briefly describes strategic attack as part of the joint fires function; JP 3-0, *Joint Operations*, Washington, D.C.: Joint Chiefs of Staff, January 17, 2017.

[19] JP 3-0, 2017, p. III-31.

force may have capabilities to conduct strategic attacks" as part of the joint fires functions.[20] Annex 3-70 does not reference this section of joint doctrine. Instead, it cites a DoD policy document, which lists strategic attack as an Air Force function.[21] Annex 3-70 seems to imply that strategic attack is an exclusively Air Force function, which is not fully consistent with joint doctrine.

Another example of Air Force doctrine that is somewhat but not fully aligned with corresponding joint doctrine is Annex 3-13, *Information Operations.*[22] Air Force information operations doctrine offers links to and considerations for joint processes, but it provides insufficient context. The 42-page Air Force annex offers few of the details compared with the four joint doctrine documents associated with information operations—JP 3-13, *Information Operations*, and its related documents.[23] In addition, such issues as the information operations assessment framework, military deception planning methodology, tenets of military deception, the military information support operations process, the operations security process, and multinational information operations are not referenced.

The Air Force's Annex 3-60, *Targeting*, and the corresponding JP 3-60, *Joint Targeting*, provide another case of largely analogous documents that are more aligned than divergent. The two documents agree on several key principles, including the need for targeting to be clearly tied to objectives and to think about all possible tools in deciding how to affect any given target.[24] Both documents also cover the differences between dynamic and deliberate targeting.[25] The annex endorses and briefly summarizes the joint targeting cycle (JTC) and refers to joint doctrine for descriptions of target characteristics.[26] Annex 3-60 also includes a diagram (see Figure 2.4) that shows how the Air Force and joint processes relate.[27]

---

[20] JP 3-0, 2017, pp. III-31. Annex 3-70 (2017) does cite JP 3-0 in another place in the annex but related to something different.

[21] DoD Directive 5100.01, *Functions of the Department of Defense and Its Major Components*, Washington, D.C.: U.S. Department of Defense, December 21, 2010.

[22] Annex 3-13, *Information Operations*, Washington, D.C.: Headquarters, U.S. Air Force, April 28, 2016, p. 12. Other examples of joint language include: Annex 3-03, *Counterland Operations*, Washington, D.C.: Headquarters, U.S. Air Force, March 17, 2017, p. 33; Annex 3-60, *Targeting*, Washington, D.C.: Headquarters, U.S. Air Force, February 14, 2017, p. 26.

[23] JP 3-13, *Information Operations*, Washington, D.C.: Joint Chiefs of Staff, November 2012, incorporating Change 1, November 2014.

[24] Annex 3-60, 2017, p. 6; JP 3-60, 2013, pp. I-7–I-8.

[25] Annex 3-60, 2017, pp. 4–5, 30–31.

[26] Annex 3-60, 2017, pp. 8–12; JP 3-60, 2013, pp. I-2–I-5.

[27] Annex 3-60, 2017, pp. 11, 55.

Figure 2.4. Relationship Between the Joint Targeting Cycle and the Air Tasking Cycle

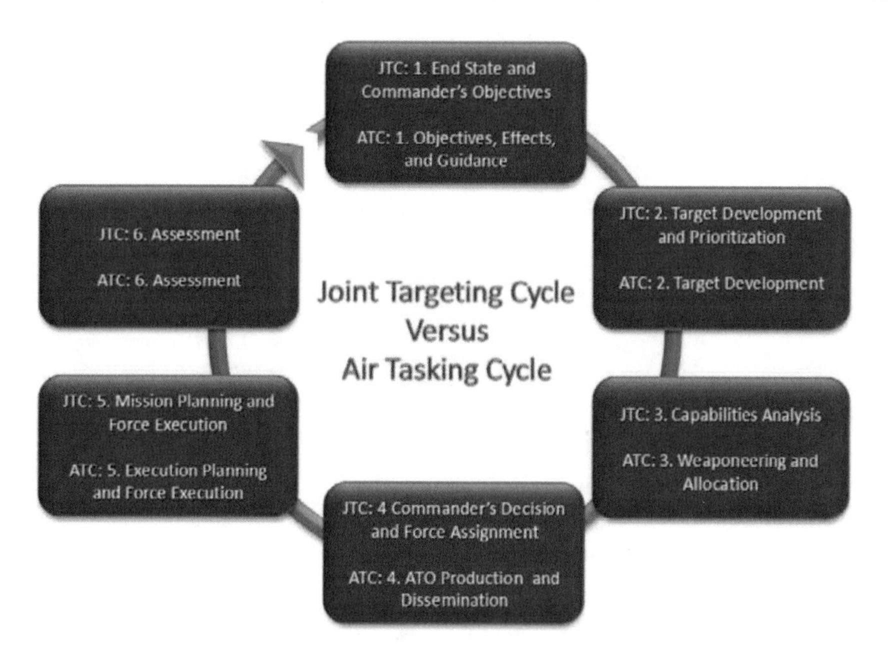

SOURCE: Annex 3-60, 2017, p. 55.
NOTE: ATC = air tasking cycle.

Despite these important similarities that contribute to alignment, the differences between ATC and JTC activities are not always clear, and the two documents describe similar activities with different terms. Air Force targeting doctrine uses the terms *kinetic* and *non-kinetic* to describe types of weapons rather than the joint terms *lethal* and *nonlethal*.[28] Additionally, joint doctrine identifies several special categories of targets, including high-value target, high-payoff target, time-sensitive target (TST), and component-critical target (CCT).[29] Air Force doctrine defines TSTs and CCTs but not the other categories.[30] Elsewhere, Annex 3-60 notes that "some Airmen" prefer the term "time-critical target" (TCT) instead of TST and explains the logic of this choice.[31] Although the document does not explicitly endorse the term TCT, it does not explicitly discourage its use either.

Many Annexes Are More Divergent from Than Aligned with Joint Doctrine

Although almost half (15 of 31) of the Air Force doctrine documents we reviewed are either significantly aligned with or more aligned with than divergent from joint doctrine, 14 are more

---

[28] *Nonlethal weapons* are defined in the *DoD Dictionary of Military and Associated Terms* (DoD, 2017). *Lethal* is not defined, but it is widely used in joint doctrine (see DoD, 2017, p. 169; Annex 3-0, *Operations and Planning*, Washington, D.C.: Headquarters, U.S. Air Force, November 4, 2016, p. 53).

[29] JP 3-60, 2013, pp. I-9.

[30] Annex 3-60, 2017, pp. 4, 45.

[31] Annex 3-60, 2017, p. 5.

divergent than aligned. Finding this number of Air Force doctrine documents, including Vol. I of the Air Force's *Core Doctrine* document,[32] that diverge in notable ways from joint doctrine suggests that there is incomplete acceptance of joint constructs in the Air Force. Our comparison methodology, described earlier in this chapter, identified several factors that can contribute to the divergence, including key service principles differing from joint principles, service doctrine not including key joint topics or covering them in insufficient detail, service doctrine not providing sufficient links to joint documents and processes, and service doctrine using different processes or terms from joint doctrine. The following sections provide examples of these factors drawn from our comparisons.

Key Air Force Constructs Differ Notably from Joint Constructs

Air Force doctrine documents describe many key constructs that differ in significant ways from analogous joint constructs. These range from different approaches to key concepts (e.g., operational design) to differences in particular terms. Also, these differences are not always well explained, and it is not always clear whether the differences are intentional. As a result, airmen may not always know which constructs are service-specific or how alternative constructs relate to constructs presented in joint doctrine.

The following sections describe some of the differences between Air Force and joint doctrine documents. The Air Force may decide that some of these differences must be preserved because of the unique characteristics of airpower or shortcomings in joint doctrine. In these cases, clearly outlining the nature of and reasons for the differences could help airmen navigate joint settings.

Mission Command and Centralized Control and Decentralized Execution

The C2 philosophy described in joint doctrine is *mission command*, a philosophy that has historical roots in land services, including the U.S. Army.[33] JP 1 defines *mission command* as the "conduct of military operations through decentralized execution based on mission type orders" and states that it is the preferred method of exercising C2 over a joint force.[34] Other joint documents, such as JP 3-0,[35] use similar language.

The Air Force has its own C2 philosophy, centralized control and decentralized execution (CCDE), which it argues is uniquely suited to airpower. Although CCDE and mission command have some similarities, they also have significant differences. Air Force doctrine contends that all airpower should be placed under the control of a single airman who directs, integrates, plans,

---

[32] LeMay Center, 2015a.

[33] James W. Harvard, "Airmen and Mission Command," *Air & Space Power Journal*, March–April 2013. Headquarters, Department of the Army, *Mission Command*, Washington, D.C., ADP 6-0, May 2012, incorporating changes as of March 12, 2014a.

[34] JP 1, 2013, pp. V-15– I-18.

[35] JP 3-0, *Joint Operations*, Washington, D.C.: Joint Chiefs of Staff, January 17, 2017.

coordinates, and assesses the use of airpower across the range of military operations (ROMO). In current doctrine, centralized control means that the COMAFFOR, through the air operations center (AOC), tasks an individual sortie with its mission, including the planning of most of the enabling details and establishing the operational constraints. Decentralized execution enables the operational unit to conduct the detailed mission planning and select the tactics necessary to accomplish the mission tasking.[36]

Although decentralized execution is intended to delegate authority to lower-level commanders in order to foster disciplined initiative and tactical flexibility, the C2 approach in current Air Force doctrine appears to be significantly more rigid than that of mission command. Air Force *Basic Doctrine* does warn that centralized control "should not become a recipe for micromanagement, stifling the initiative subordinates need to deal with combat's inevitable uncertainties."[37] However, the Air Force's preferred way of directing air operations is through the ATC managed by the AOC, which leaves tactical-level units with little leeway for the planning of operations.[38] The resulting ATO includes weaponeering and allocation decisions that can be quite directive. Weaponeering includes, among other things, target identification and description, recommended aim points, desired scope, level(s) and duration of damage, weapon systems and munition recommendations, and potential fusing requirements. Allocation includes decisions on the total number of sorties or missions by weapon system type allocated to each objective or task.[39]

Both C2 approaches are similar in that they intend to give tactical commanders the flexibility to respond to unfolding conditions during the execution of a tactical mission. They differ in that mission command is predicated on the principle that each echelon should have a great deal of flexibility in planning its operations via mission-type orders. CCDE, however, is predicated on the importance of centralizing control under a single, generally theater-level, headquarters that is responsible for most operational planning and decisions.

Moreover, Air Force doctrine does not clearly explain mission command or how it differs from CCDE. Neither *Basic Doctrine*, which was last revised in February 2015, nor the Air Force glossary defines or uses the term *mission command*.[40] However, the adoption of *mission command* by the joint

---

[36] LeMay Center, 2015a, pp. 67–69. Doctrine developers have argued that the more general principle of CCDE, rather than its current manifestation, could accommodate arrangements with less centralized control than has been practiced in recent years; Brian McLean, "Reshaping Centralized Control/Decentralized Execution for the Emerging Operating Environment," *Over the Horizon*, March 13, 2017; Jeffrey Hukill et al., *Air Force Command and Control: The Need for Increased Adaptability*, Maxwell Air Force Base, Ala.: Air Force Research Institute, Air Force Research Institute Papers 2012-5, July 2012.

[37] LeMay Center, 2015a, p. 68.

[38] LeMay Center, 2016, p. 33. The close relationship between the ATC and the JTC is described in Annex 3-60, 2017.

[39] Annex 3-60, 2017, pp. 70–75.

[40] LeMay Center, 2015a; Headquarters, U.S. Air Force, *Air Force Glossary*, Washington, D.C., 2016.

community is not new, because it is in the August 2011 edition of JP 3-0 and the March 2013 edition of JP 1. This service-level C2 approach may put Air Force officers at a disadvantage when they are attached to joint planning staffs both because it can leave them unfamiliar with the joint concept of mission command and because it can deny them the opportunity to participate in broader planning processes when they are junior and midgrade officers.

Cyberspace

Air Force cyberspace doctrine, contained in Annex 3-12, provides another example of differing constructs. Although Annex 3-12 uses and cites the definition of cyberspace presented in JP 3-12 (R),[41] the documents do not share a common framework for cyberspace. Joint doctrine describes cyberspace in terms of three layers (the physical network, logical network, and cyber-persona).[42] Air Force doctrine instead begins by distinguishing cyberspace from information operations and describes the operational environment and infrastructure relationships based on non-DoD documents rather than joint doctrine.[43]

Annex 3-12 does share some common threat categories and descriptions with JP 3-12(R). Several threat terms and definitions are verbatim from those used in JP 3-12 (R), but JP 3-12 (R) cites the 2006 National Military Strategy for Cyberspace Operations instead.[44] The annex also contains at least four threat categories not included in JP 3-12 (R): catastrophic threat, disruptive threat, natural threat, and accidental threat.[45] Annex 3-12 still references the global information grid rather than using the 2015 term: DoD information network. Annex 3-12 also lacks references to joint operating terms, such as *joint network operations control center*, *joint cyberspace center* (JCC), and *defensive cyberspace operations and internal defensive measures*. It is worth noting that the annex has not been updated since 2011. This could explain some of the substantive gaps and inconsistencies with joint doctrine.

Air and Missile Defense

An instance of an alternative Air Force construct that is intentionally competitive with a joint construct can be found in the 2015 version of Annex 3-01, *Counterair Operations.*[46] Some of the

---

[41] Annex 3-12, 2011, p. 2; JP 3-12 (R), *Cyberspace Operations*, Washington, D.C.: Joint Chiefs of Staff, February 5, 2013, p. GL-4. The (R) indicates that this is a redacted version of the full JP.

[42] JP 3-12 (R), 2013, p. I-3.

[43] The annex references a homeland security document in its discussion of cyber infrastructure; Annex 3-12, 2011, pp. 3–5. The annex refers readers desiring more information on "physical, syntactic, and semantic layers of cyberspace" to Chapter 10 of Martin C. Libicki, *Conquest in Cyberspace: National Security and Information Warfare*, New York: Cambridge University Press, 2007.

[44] Annex 3-12, 2011, pp. 13–16.

[45] Annex 3-12, 2011, p. 15.

[46] Annex 3-01, *Counterair Operations*, Washington, D.C.: Headquarters, U.S. Air Force, October 27, 2015.

differences between this document and JP 3-01, *Countering Air and Missile Threats*, are expected to be addressed in the next version of Annex 3-01.[47] Still, we discuss the differences here to highlight how using Air Force doctrine to advocate for changes to joint doctrine can create potentially confusing divergences between the two.

JP 3-01 provides doctrine for countering both air and missile threats. The joint conception of countering air and missile threats consists of a "combination of counterair and integrated air and missile defense (IAMD). Counterair is the foundational framework at the theater level. IAMD is an approach that synchronizes aspects of counterair with global missile defense (MD); homeland defense (HD); global strike; and counter-rocket, artillery, and mortar (C-RAM)."[48] Furthermore, IAMD is defined as

> the integration of capabilities and overlapping operations to defend the homeland and US national interests, protect the joint force, and enable freedom of action by negating an enemy's ability to create adverse effects from their air and missile capabilities. IAMD incorporates offensive and defensive measures to create a comprehensive joint and combined force capable of preventing an enemy from effectively employing its offensive air and missile weapons. IAMD is designed to deter, and failing that, to prevent an enemy from effectively employing air and missile assets.[49]

JP 3-01 further states that "at the theater level, IAMD consists of [defensive counterair] supported by offensive counterair (OCA) attack operations. Beyond the theater level, IAMD emphasizes the integration of these counterair operations with global MD, HD, and global strike. IAMD also includes C-RAM at the tactical level."[50]

Annex 3-01's brief discussion of IAMD focuses on its on theater-level elements that are included in the counterair framework. It largely ignores the strategic or national-level components of IAMD.[51] In addition, while the Air Force sees IAMD as a subset of activities within the broader counterair framework, joint doctrine sees it as a separate activity focused on broader national objectives. The Air Force did not accept the joint community's initial IAMD concept because it believed that it had the potential to violate the principle of unity of command by interfering with the joint forces air component commander's (JFACC's) ability to determine

---

[47] JP 3-01, *Countering Air and Missile Threats*, Washington, D.C.: Joint Chiefs of Staff, April 21, 2017.

[48] JP 3-01, 2017, p. I-1.

[49] JP 3-01, 2017, pp. I-10–I-11.

[50] JP 3-01, 2017, p. I-11.

[51] Annex 3-01, 2015, p. 25.

the balance between OCA and defensive counterair capabilities conducted in support of the counterair mission.[52] Annex 3-01 thus states that, from

> an Airman's perspective, the IAMD model carries the potential to split activities between offense and defense, which, from an Airman's perspective, may fracture unity of command and unity of effort. Thus, Airmen should always advocate the counterair framework vice IAMD when discussing countering air and missile threats, even in a joint context.[53]

Figure 2.5 shows the two competing frameworks as presented in JP 3-01 and Annex 3-01.

Figure 2.5. Joint and Air Force Counterair and IAMD Frameworks

The Relationship Between Counterair and Integrated Air and Missile Defense
Counterair Mission
Offensive Counterair
• Attack Operations
• Suppression of Enemy Air Defenses
• Fighter Escort
• Fighter Sweep
Defensive Counterair
• Active Air and Missile Defense
• Passive Air and Missile Defense
• Homeland Defense
• Global Missile Defense
• Global Strike
• Counter Rocket, Artillery, and Mortar
Integrated Air and Missile Defense Approach

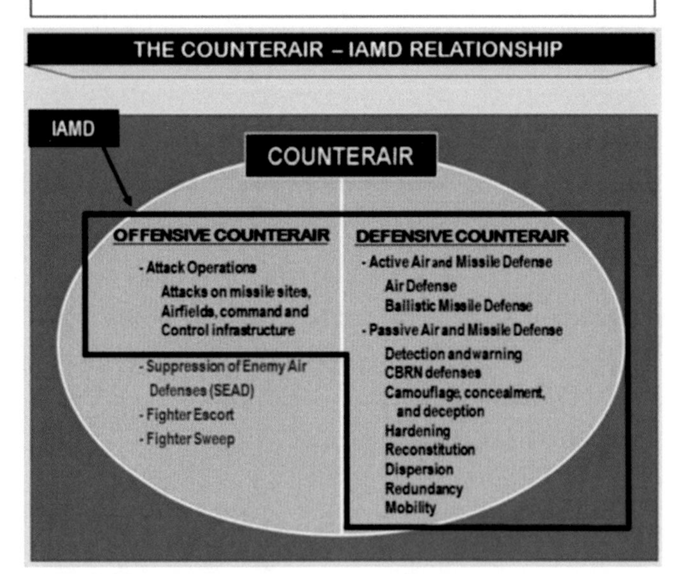

SOURCES: JP 3-01, 2017, p. I-12; Annex 3-01, 2015, p. 24.

[52] Some of the key differences between the Air Force and the joint community regarding IAMD were resolved during the writing of the 2017 edition of JP 3-01. The next edition of Annex 3-01 is expected to more closely reflect JP 3-01 in regard to IAMD; LeMay Center, 2017c.

[53] Annex 3-01, 2015, p. 24.

Intelligence Principles

Air Force and joint doctrine list different intelligence principles. In Figure 2.6, the list in purple indicates principles that are discussed in joint doctrine but not in Air Force doctrine. These principles relate to the process for developing finished intelligence. For example, JP 2-0 discusses the importance of perspective, meaning that the analyst should think like the enemy and understand its culture and social structures. JP 2-0 also encourages analysts to maintain intellectual integrity by taking steps to avoid presenting information that simply confirms the analyst's or the requester's biases or preferences.[54] Joint doctrine sets standards for communicating the level of confidence an analyst has in an analytic judgment.[55] The joint principle of excellence partially overlaps with the Air Force principles of accurate, timely, relevant, accessible, and secure.[56]

Figure 2.6. Air Force and Joint Intelligence Principles

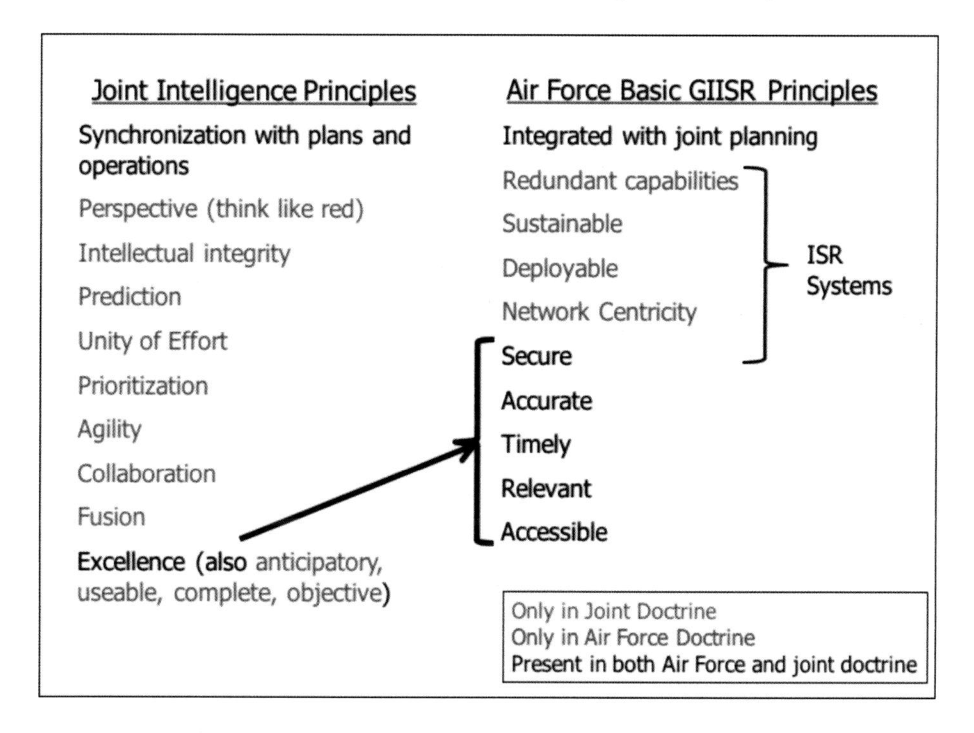

SOURCES: JP 2-0, 2013, pp. II-1–II-2; Annex 2-0, *Global Integrated Intelligence, Surveillance & Reconnaissance Operations*, Washington, D.C.: Headquarters, U.S. Air Force, January 29, 2015, pp. 6–8.
NOTE: GIISR = global integrated intelligence, surveillance, and reconnaissance; ISR = intelligence, surveillance, and reconnaissance.

---

[54] JP 2-0, *Joint Intelligence*, Washington, D.C.: Joint Chiefs of Staff, October 22, 2013, pp. II-1–II-2.

[55] JP 2-0, 2013, pp. A-1–A-2.

[56] The joint principle of excellence covers those principles as well as anticipatory, useable, complete, and objective; JP 2-0, 2013, pp. II-1–II-2.

The Air Force has several intelligence principles that are not listed in joint doctrine, as indicated in blue. These are related to ISR systems rather than the broader intelligence process: redundant capabilities, sustainable, deployable, and network centricity.[57]

In some instances, Air Force doctrine uses terms that are different from those used in joint doctrine. Because the terms are not always defined, it is not clear whether the terms are different and, if so, whether that difference is intentional. Annex 2-0, *Global Integrated Intelligence, Surveillance & Reconnaissance Operations*, for example, does not define the key joint term *intelligence* or explain its relationship with the Air Force term GIISR. The Air Force glossary does present the joint definition of *intelligence*:

> 1. The product resulting from the collection, processing, integration, evaluation, analysis, and interpretation of available information concerning foreign nations, hostile or potentially hostile forces or elements, or areas of actual or potential operations. 2. The activities that result in the product. 3. The organizations engaged in such activities.[58]

However, Annex 2-0 does not contain this information. In some cases, the annex uses the term GIISR to refer to intelligence activities and organizations. In other cases, GIISR seems to more closely relate to the joint concept of ISR, one of many intelligence activities that is defined in both Air Force and joint doctrine as

> an activity that synchronizes and integrates the planning and operations of sensors, assets, and processing, exploitation, and dissemination systems in direct support of current and future operations.[59]

Annex 2-0 defines GIISR as

> cross-domain synchronization and integration of the planning and operation of ISR assets; sensors; processing, exploitation and dissemination systems; and, analysis and production capabilities across the globe to enable current and future operations.[60]

This definition largely overlaps with ISR, but it also mentions analysis and production capabilities that are part of the joint concept of intelligence.[61] Without a clear explanation, it is difficult for the reader to understand the distinctions.

Moreover, the document refers to combatant commands having a GIISR process, which contains most of the same steps as the joint intelligence process: planning and direction, collection, processing and exploitation, analysis and production, and dissemination and

---

[57] Annex 2-0, 2015, pp. 6–8.

[58] Headquarters, U.S. Air Force, 2016, p. 46; DoD, 2017, p. 116.

[59] DoD, 2017, p. 118.

[60] Annex 2-0, 2015, p. 2.

[61] Annex 2-0, 2015, p. 2.

integration. Joint doctrine also includes evaluation and feedback.[62] GIISR methodologies and products largely overlap with joint categories of intelligence. The two lists share several categories, such as warning, current, general military, target, and scientific and technical.[63]

Operational Design

Both Air Force and joint doctrine use the language of operational design in discussing operational planning and identify similar steps in the operational design process. However, they differ in how they describe and illustrate the operational design process.[64] An important step in the operational design process is the development of an operational approach that describes the operational environment and the commander's approach to "achieving the desired end state."[65]

The joint conception of the operational approach uses outcome-oriented lines of effort (LOEs) to move from current conditions to a desired end state, as shown in Figure 2.7. Each LOE consists of a series of tasks or missions. An associated defeat or stability mechanism is designed to support the achievement of several objectives—which, in turn, create conditions that enable the achievement of the desired end state. By identifying the relevant defeat and stability mechanisms, the commander explains how he or she wants to create effects along a line of operation or LOE.[66]

---

[62] JP 2-0, 2013, pp. I-5, I-21; Annex 2-0, 2015, p. 45.

[63] Joint doctrine also includes counterintelligence, estimative intelligence, and identity intelligence. In a minor terminological difference, Air Force doctrine uses the term *indications and warning*, and joint doctrine uses *warning intelligence*. Annex 2-0, 2015, p. 49; JP 2-0, 2013, p. I-18.

[64] These differences occur despite the fact that Annex 3-0 provides embedded links that lead directly to JP 5-0's description of the operational design process.

[65] JP 5-0, *Joint Operation Planning*, Washington, D.C.: Joint Chiefs of Staff, August 11, 2011, p. III-13.

[66] JP 5-0, 2011, pp. III-15, III-16, III-29–III-30.

Figure 2.7. Sample Joint Operational Approach

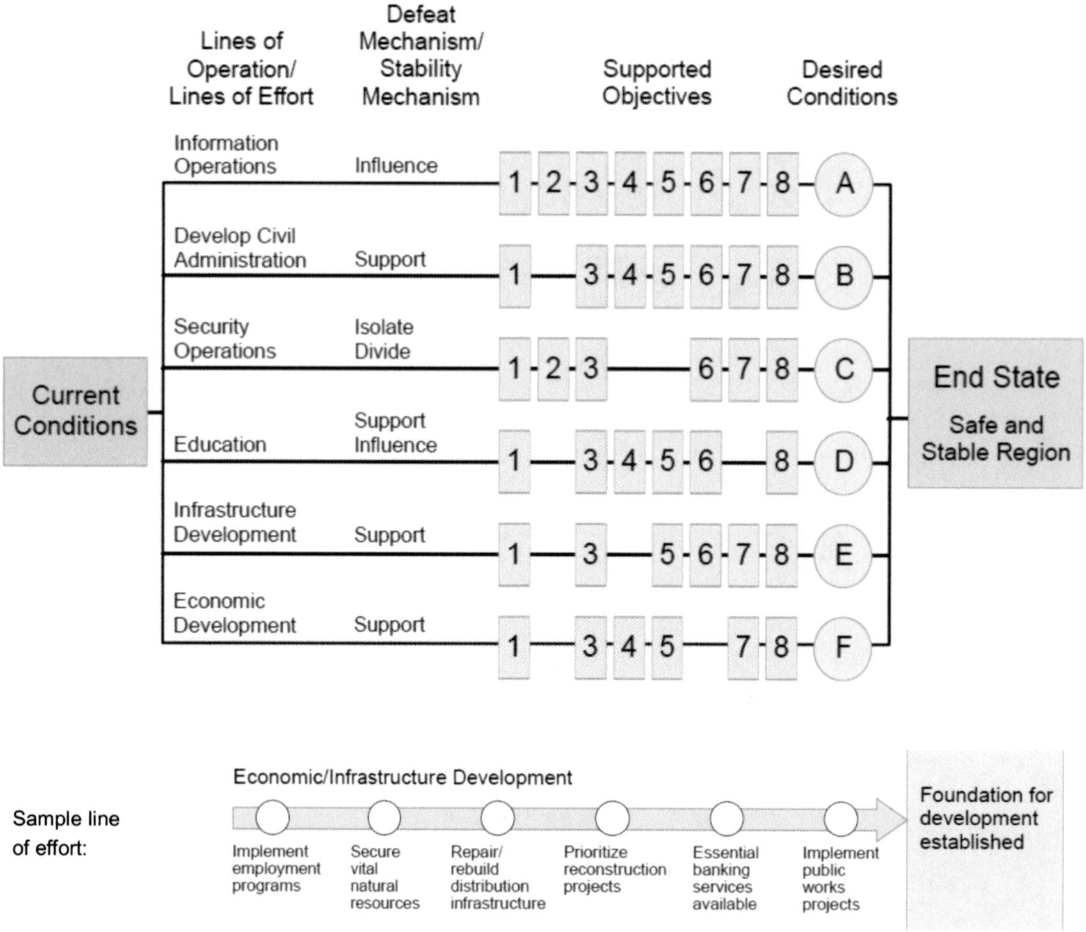

Lines of Operation/ Lines of Effort
Defeat Mechanism/ Stability Mechanism
Supported Objectives
Desired Conditions
Current Conditions
Information Operations — Influence — 1-2-3-4-5-6-7-8 — A
Develop Civil Administration — Support — 1—3-4-5-6-7-8 — B
Security Operations — Isolate Divide — 1-2-3—6-7-8 — C
Education — Support Influence — 1—3-4-5-6—8 — D
Infrastructure Development — Support — 1—3—5-6-7-8 — E
Economic Development — Support — 1—3-4-5—7-8 — F
End State
Safe and Stable Region

Sample line of effort:
Economic/Infrastructure Development
Implement employment programs
Secure vital natural resources
Repair/ rebuild distribution infrastructure
Prioritize reconstruction projects
Essential banking services available
Implement public works projects
Foundation for development established

SOURCE: JP 5-0, 2011, pp. III-15, III-29.

The Air Force's illustration of operational approach, shown in Figure 2.8, begins with tactical, operational, or strategic actions rather than current conditions. These actions create effects along functional LOEs, which are associated with a center of gravity, the commander's objectives, and the operation's end state. Joint doctrine warns against using functional LOEs, such as diplomacy or economics, as this can result in agency stovepipes during the execution of an operation.[67] The Air Force description also omits stability and defeat mechanisms, which are not discussed in Annex 3-0.

[67] JP 5-0, 2011, p. III-28.

Figure 2.8. Sample Air Force Operational Approach

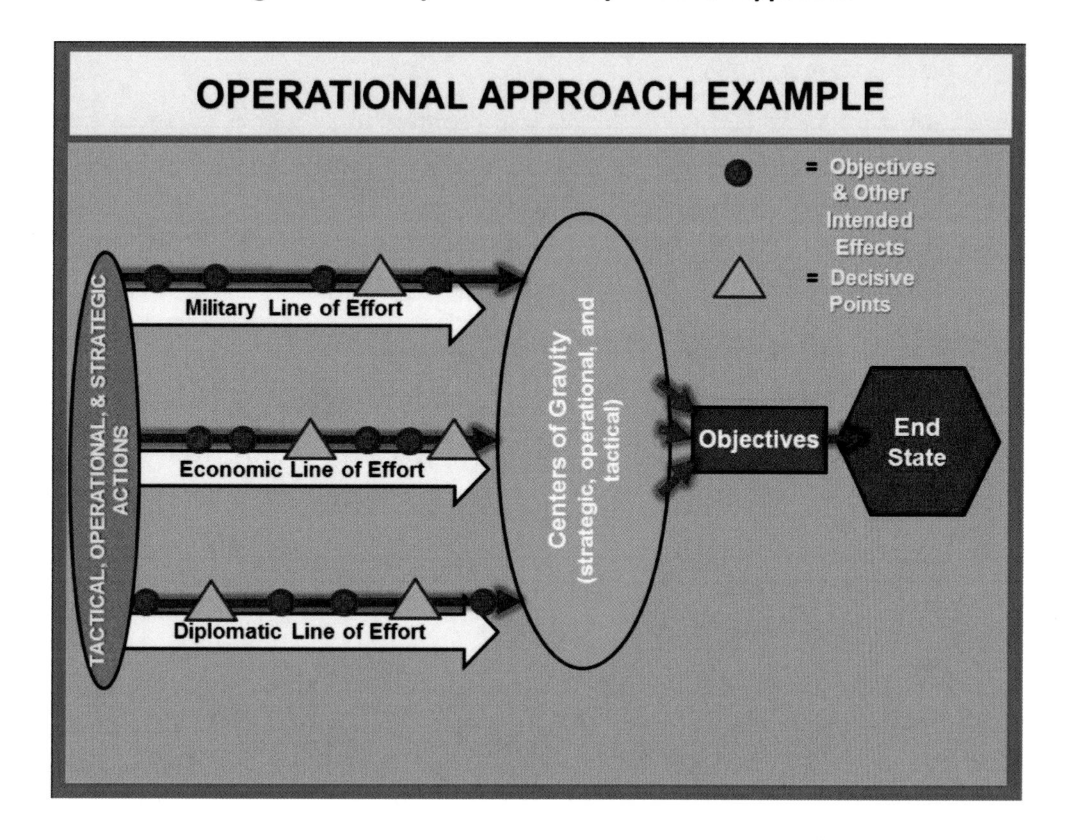

SOURCE: Annex 3-0, 2016, p. 50.

Air Force and joint doctrine also have different ways of describing the tasks involved in defining (or framing) the problem during the operational design process. The joint description of defining the problem includes the following tasks:

- Define and articulate the tensions between current conditions and desired conditions at the end state.
- Define and articulate the tension elements within the operational environment, which may change or remain the same to achieve desired end states.
- Define and articulate the opportunities and threats that either can be exploited or will impede the JFC from achieving the desired end states.
- Define and articulate the limitations on the JFC's freedom of action.[68]

These tasks should result in a concise problem statement that clearly defines the problem to be solved.

By way of contrast, the Air Force identifies the following tasks as part of the define-the-problem step:

- Determine the strategic context and systemic nature of the problem(s).
- Synthesize strategic guidance.

---

[68] JP 5-0, 2011, pp. III-12–III-13.

- Identify strategic trends.
- Identify gaps in knowledge and assumptions about the problem(s).
- Identify the operational problem(s).[69]

Relatedly, Annex 3-0 introduces the idea of the coercion continuum, which is meant to explain how military action ranges from "pure threat" to "pure force." Although this concept can clarify the many types of warfare and their corresponding levels of intensity, it is not present in joint doctrine. Moreover, there is no explanation of how this continuum relates to joint planning constructs.[70]

Overall, although Air Force and joint doctrine discuss operational design, their descriptions are different in many important ways.

Air Force Doctrine Overlooks Key Joint Topics or Covers Them in Insufficient Detail

Omitting key joint topics or describing them in insufficient detail is one of the major factors causing Air Force doctrine to diverge from joint doctrine. Possibly the most noteworthy example is found in *Core Doctrine,* Vol. I, *Basic Doctrine*, the Air Force's capstone doctrinal document.[71] This document provides the Air Force's basic beliefs about the application of Air Force capabilities across ROMO, provides guidance on the proper employment of airpower, sets the foundation for educating airmen about airpower, and guides the development of all other Air Force doctrine. *Basic Doctrine* does not include a discussion of several topics that are central to the joint capstone doctrinal document (JP 1) or *Joint Operations* (JP 3-0). Some of the key joint constructs that *Basic Doctrine* omits are unified action, the joint functions, the phasing of joint operations, stability operations and activities, and the balance between offense, defense, and stability activities.

Unified action is an important joint concept that is extensively discussed and used in both JP 1 and JP 3-0. *Unified action* is defined as "synchronization, coordination, and/or integration of the activities of governmental and nongovernmental entities with military operations to achieve unity of effort."[72] JP 3-0 devotes roughly four pages to a discussion of unified action, but it is not mentioned in *Basic Doctrine.*

Joint functions are another important joint concept discussed extensively in JP 1 and JP 3-0. *Joint functions* are defined as "related capabilities and activities placed into six basic groups of C2, intelligence, fires, movement and maneuver, protection, and sustainment to help joint force commanders synchronize, integrate, and direct joint operations."[73] JP 1 notes that these

---

[69] Annex 3-0, 2016, p. 48.

[70] Annex 3-0, 2016.

[71] LeMay Center, 2015a.

[72] JP 1, 2013, p. GL-12.

[73] JP 3-0, 2017, p. GL-11.

groupings "facilitate planning and employment of the joint force."[74] Similarly, JP 3-0, which devotes an entire chapter to describing them, notes that the "joint functions reinforce and complement one another, and integration across the functions is essential to mission accomplishment."[75] *Basic Doctrine* does not discuss the joint functions or provide links to information about them.[76] The importance and prevalence of these functions, as shown in their coverage in JP 1, JP 3-0, and other joint documents, suggests that airmen will need to understand them to function effectively in joint staff and leadership positions and that not including them in Air Force doctrine was a choice rather than an oversight.[77]

The gaps in Air Force doctrine are also apparent in several operational annexes, such as Annex 2-0.[78] This annex covers many of the topics in the joint capstone document on intelligence (JP 2-0), but in much less detail. The differing levels of detail are suggested by the comparative lengths—Annex 2-0 is 62 pages, while JP 2-0 runs 144 pages. Joint intelligence doctrine also includes three other detailed documents that have little coverage in Annex 2-0.[79] As discussed with the example of Annex 3-27, *Homeland Operations*, it is possible to write a brief annex that still introduces airmen to all of the key joint constructs.[80] However, in the case of Annex 2-0, the short document leaves out key information. As a specific example, Annex 2-0 briefly mentions the joint intelligence preparation of the operating environment (JIPOE) process and, by way of explanation, draws a comparison with service-level intelligence preparation of the battlespace.[81] However, the annex does not explain either process or refer the reader to the discussion of the JIPOE in JP 2-0 or to JP 2-01.3, the document dedicated to this topic.[82]

---

[74] JP 1, 2013, p. I-17.

[75] JP 3-0, 2017, pp. III-1.

[76] LeMay Center, 2015a.

[77] The functions are also used as a framework in Joint Chiefs of Staff, *Capstone Concept for Joint Operations: Joint Force 2020*, Washington, D.C.: U.S. Department of Defense, September 10, 2012, pp. 8–13.

[78] Annex 2-0, 2015.

[79] JP 2-0, 2013; JP 2-01, *Joint and National Intelligence Support to Military Operations*, Washington, D.C.: Joint Chiefs of Staff, January 5, 2012; JP 2-03, *Geospatial Intelligence in Joint Operations*, Washington, D.C.: Joint Chiefs of Staff, October 31, 2012; JP 2-01.3, *Joint Intelligence Preparation of the Operational Environment*, Washington, D.C.: Joint Chiefs of Staff, May 21, 2014. JP 2-01.2, *Counterintelligence and Human Intelligence in Joint Operations*, is not available to the public and not included in this analysis. Annex 2-0 (2015) does refer the reader to JP 2-01, *Joint and National Intelligence Support to Military Operations*, for more information on how the Air Force distributed common ground system (DCGS) integrates into the joint intelligence process. The annex uses and cites definitions from JP 2-03 but does not refer the reader to that document or discuss it in further detail.

[80] Annex 3-27, 2016.

[81] Annex 2-0, 2015, p. 14.

[82] JP 2-0, 2013; JP 2-01.3, 2014.

As another example, Annex 3-0, *Operations and Planning*, does not cover some important joint operational concepts discussed in both JP 3-0 and the 2011 version of JP 5-0, including course-of-action development, stability operations, and the joint operations model. Furthermore, Annex 3-0 lacks a discussion of using the political, military economic, social, information, and infrastructure (PMESII) framework to characterize the operational environment as part of the operational design process.[83]

Planning Processes and Considerations Are Often Missing from Air Force Doctrine

Several Air Force doctrine documents lack discussion of or references to the joint planning process, which was called the joint operation planning process (JOPP) at the time this analysis was conducted. This process is central to joint military operations and, therefore, must be understood to operate effectively as a joint leader. However, the qualitative comparison revealed that substantive discussions of planning were less prevalent and detailed in Air Force documents than in joint doctrinal documents.

JP 3-01, *Countering Air and Missile Threats*, for example, has an extensive discussion the planning processes involved in air and missile defense.[84] However, Annex 3-01 largely ignores this issue.[85] JP 3-34, *Joint Engineer Operations*, includes a discussion of engineering considerations at each stage of the JOPP.[86] Air Force doctrine discusses issues that airmen need to consider during deliberate and crisis planning but does not methodically link them to the JOPP or other planning process.[87] Annex 3-34 does not refer the reader to joint doctrine for more information on any of these topics. Annex 3-22, *Foreign Internal Defense*, also lacks a detailed discussion of planning, while JP 3-22 devotes an entire 20-page chapter to the topic.[88]

Similarly, Annex 3-2, *Irregular Warfare*, spends six pages on irregular warfare strategy and planning considerations but does not discuss nor provide direct links to the joint planning processes found in joint irregular warfare doctrine. The annex notes that joint doctrine has different planning constructs along the ROMO, so "the Air Force component, the [irregular warfare] planner should mirror the planning construct used by the respective combatant

---

[83] JP 5-0, 2011; Annex 3-0, 2016; JP 3-0, 2017. A revised edition of JP 5-0, Joint Planning, was released in June 2017 after this analysis was completed; as a result, the 2011 version of JP 5-0, Joint Operation Planning, is the focus of this comparison.

[84] JP 3-01, 2017.

[85] Annex 3-01, 2015.

[86] JP 3-34, *Joint Engineer Operations*, Washington, D.C.: Joint Chiefs of Staff, January 6, 2016, pp. IV-2–IV-9.

[87] Annex 3-34, *Engineer Operations*, Washington, D.C.: Headquarters, U.S. Air Force, December 30, 2014, pp. 28–31.

[88] Annex 3-22, *Foreign Internal Defense*, Washington, D.C.: Headquarters, U.S. Air Force, July 2015; JP 3-22, *Foreign Internal Defense*, Washington, D.C.: Joint Chiefs of Staff, July 12, 2010.

command, either steady-state or contingency."[89] However, it does not refer the reader to extensive discussion of these potential planning processes in joint doctrine, including

- JP 3-24's discussion of the design principles for planning counterinsurgency operations
- JP 5-0's examples of planning for stability and counterinsurgency operations
- Chapter IV in JP 3-07, which discusses stabilization planning
- Chapter IV in JP 3-26, which discusses counterterrorism planning
- Chapter IV in JP 3-22, which discusses foreign internal defense (FID) planning
- Chapter III in JP 3-20, which discusses security cooperation planning
- Chapter V in JP 3-05.1, which discusses unconventional warfare planning.

Air Force Doctrine Does Not Include Sufficient Linkages to Joint Documents and Processes

Air Force doctrine documents often do not include linkages to joint documents and processes. For example, Air Force doctrine on force protection describes a risk management approach that could inform a force protection planning process. However, it does not explain how this process relates to force protection in the joint planning processes.[90] Similarly, Air Force electronic warfare (EW) doctrine presents a list of EW planning responsibilities, but it does not describe how those responsibilities relate to either the JOPP or JOPP for air.[91]

Difficulties in tracking linkages to joint doctrine documents and processes are exacerbated by significant differences in organization between Air Force and joint doctrine. For example, Figure 2.9 shows the differences in organization between the Air Force and joint intelligence doctrine documents. The organization of Annex 2-0 does not parallel the organization of JP 2-0 or offer clear linkages to relevant sections of joint intelligence.[92] The left-hand column shows Annex 2-0's catalog of topics. The arrows show where each of these topics are discussed in JP 2-0. The differences in organization make it difficult for the reader to identify these connections.

---

[89] Annex 3-2, *Irregular Warfare*, Washington, D.C.: Headquarters, U.S. Air Force, July 12, 2016, p. 2.

[90] Annex 3-10, *Force Protection*, Washington, D.C.: Headquarters, U.S. Air Force, April 17, 2017, pp. 27–30.

[91] Annex 3-51, *Electronic Warfare*, Washington, D.C.: Headquarters, U.S. Air Force, October 10, 2014, pp. 20–21.

[92] Annex 2-0, 2015; JP 2-0, 2013.

Figure 2.9. Differences in Air Force and Joint Intelligence Doctrine Organization

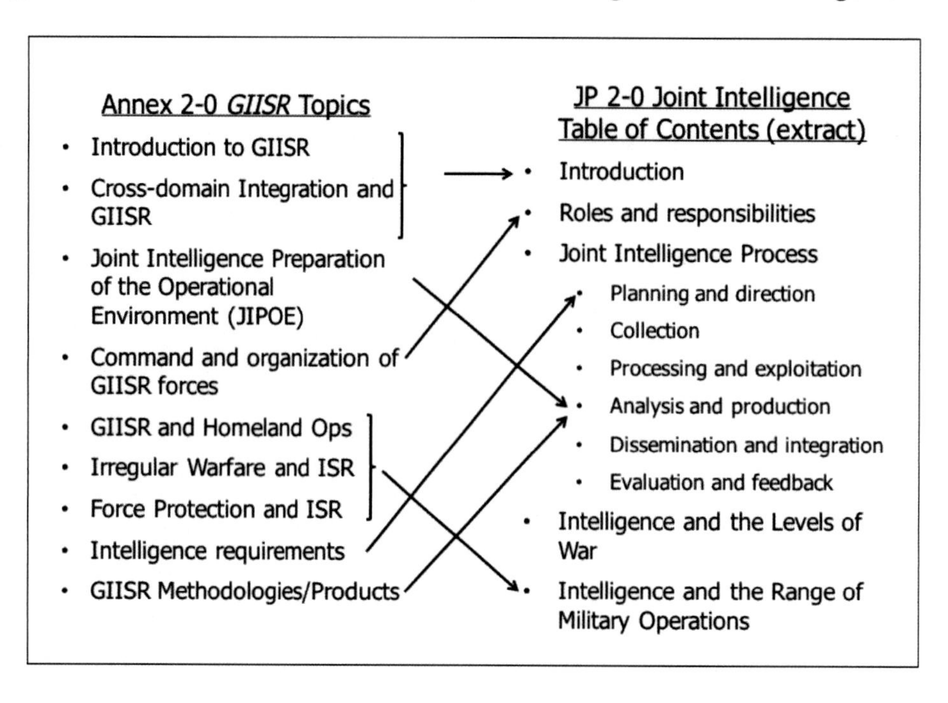

SOURCES: Annex 2-0, 2015; JP 2-0, 2013.

Generally, Air Force doctrine documents are not as well organized as joint doctrine documents. Joint doctrine developers acknowledge that most people do not read entire doctrine documents front to back. Still, they write doctrine as a book so that it has a coherent internal logic that connects all the chapters, sections, and subsections together.[93] Joint documents have a nested outline structure with big-picture context provided up front and supporting sections and subsections detailing the parts of a larger process or concept.

Air Force documents, in contrast, generally do not use a nested framework. Instead, Air Force doctrine documents are made up of a series of DTMs. Therefore, the documents read like a list of subjects, rather than a clear progression of topics, subtopics, and supporting materials. Reinforcing this structure, Air Force doctrine documents begin with a catalog of doctrine topics, rather than a table of contents, as is found in joint doctrine.

The organizational structure is, in part, because of practical considerations. The LeMay Center can use fewer resources and revise doctrine more quickly using its DTM approach. When a piece of doctrine is up for its two-year review, the LeMay Center requests input from the relevant community on which DTMs need to be revised. The center then sends only the nominated DTMs out for line editing by interested Air Force organizations, rather than revising the entire document. Although this approach has certain merits, the net effect is that it is often very difficult to understand the internal logic of these documents or how pieces of Air Force doctrine relate to joint doctrine.

[93] J-7, 2017.

This difference in organizational structures can be seen in Air Force and joint logistics doctrine. Air Force logistics DTMs are topical sections without clear transitions from subject to subject.[94] The doctrine does not explain the order of topics presented. Because of this list-type structure, Air Force doctrine touches on topics multiple times throughout the document, rather than discussing them within dedicated sections.[95] Conversely, joint logistics doctrine is organized by broad concepts or themes, with each chapter composed of clearly outlined and delineated sections, subsections, and supporting points.[96] For example, Chapter II of JP 4-0 is titled "Core Logistics Functions." The function of supply falls under this chapter and has three sections: supply chain, supply chain management, and supply chain areas. Within the supply chain area section, the document covers three additional subsections: management of supplies and equipment, inventory management, and management of global supplier networks.[97] The joint doctrine's thematic organizational structure allows for information on one subject to be contained within one area of the document, rather than being splintered throughout multiple sections. In comparison, the Air Force includes information on supply and resupply operations in at least five different sections: receiving and bedding down; supply mission; forces and infrastructure; sustainment C2; sustainment resupply, distribution, and delivery; and total asset visibility.[98]

Air Force Acceptance of a Joint Mindset Is Uneven

Our second major finding is that Air Force acceptance of a joint mindset, as judged by the tone of Air Force doctrine, is uneven. Figure 2.10 depicts the summary of assessments of Air Force doctrine's tone. Some Air Force doctrine displays a joint tone, effectively incorporating joint context and describing the service as a unique, but not superior, contributor to joint teams. Other Air Force doctrine, however, reflects a service-centric mindset. This second finding is relevant because, even if Air Force doctrine universally used joint constructs, it could still impart a mindset that fostered the impression that the Air Force does not need to integrate with other services. The next three sections present examples of Air Force doctrine that is joint in tone, more joint than service-centric, and more service-centric than joint.

---

[94] Annex 4-0, 2015, p. 1.

[95] Annex 4-0, 2015.

[96] JP 4-0, *Joint Logistics*, Washington, D.C.: Joint Chiefs of Staff, October 16, 2013, pp. v–vi.

[97] JP 4-0, 2013, pp. II-3–II-6.

[98] Annex 4-0, 2015, pp. 21–22, 58, 60, 63–64.

Figure 2.10. Air Force Doctrine Tone

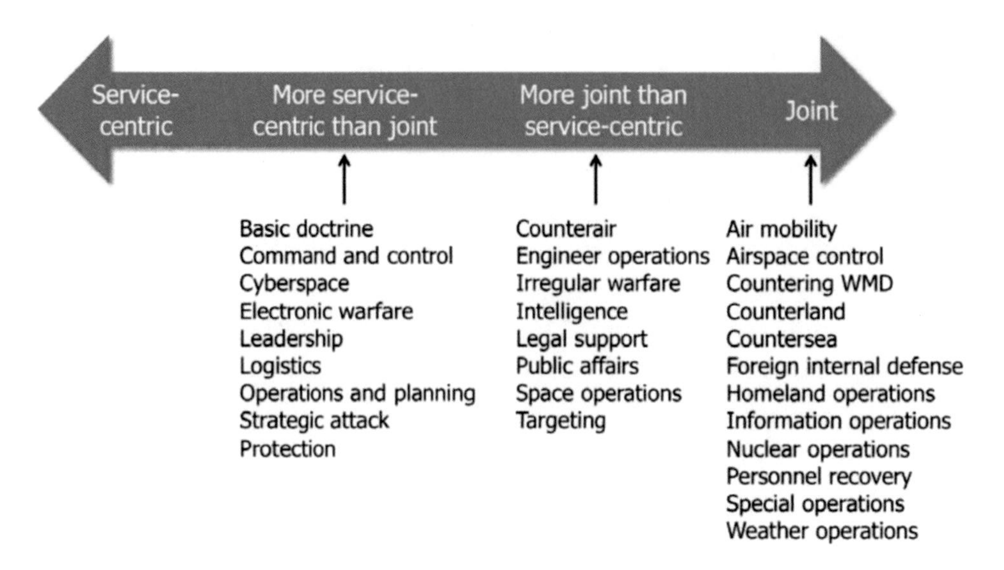

SOURCES: Annex 4-0, 2015; Annex 4-02, 2015.
NOTE: We reviewed a total of 31 Air Force doctrine documents. The two logistics annexes, 4-0 and 4-02, are combined in this figure.

Many Air Force Annexes Are Joint in Tone

Twelve of the 31 Air Force doctrine documents we examined displayed a highly joint tone, effectively incorporating joint context and describing the service as a unique, but not superior, contributor to joint teams. Air Force annexes on special operations, air mobility, foreign internal defense, and counterland operations all had a joint tone.

Annex 3-05, *Special Operations*, is an excellent example of Air Force doctrine that is joint in tone.[99] Annex 3-05 does not reference the superiority of Air Force Special Operations Command (AFSOC) over other service special operations forces (SOF) organizations. Instead, the document strikes a balanced approach by first introducing concepts in terms of their broader joint context and then discussing the same concepts as they apply to Air Force special operations forces (AFSOF). Annex 3-05 makes only one reference to "Airmen," and instead refers to the "SOF operator," reinforcing the joint nature of AFSOF.[100] Annex 3-05 predominantly describes Air Force special operations in the context of a joint setting. Annex 3-05 focuses on AFSOF's role in support of the JFC; discusses how airpower contributes to the joint mission, the JFC's campaign, and USSOCOM core activities; and explains how AFSOF integrates into joint force plans.[101]

---

[99] JP 3-05's lead agent is U.S. Special Operations Command (USSOCOM); JP 3-05, 2014; Annex 3-05, 2017.

[100] Annex 3-05, 2017, p. 16.

[101] Annex 3-05, 2017, p. 26.

Additionally, Annex 3-05 makes reference to the inherently joint nature of special operations when it notes that it "is highly unlikely [special operations] would be conducted as a single Service operation; therefore, [special operations] planning should consider joint support and coordination, to include Air Force forces."[102] Likewise, Annex 3-05 also notes the broader context in which AFSOF operate and the frequency with which AFSOF operate with not only other services but also other U.S. government agencies and multinational forces.

Another example of joint tone is Annex 3-17, *Air Mobility Operations*, which discusses air mobility in support of the joint force commander and Army and Navy roles as part of the broader transportation network.[103] The document contains a very service-specific discussion of processes and air mobility assets without being service-centric in tone.

Although Air Force foreign internal defense doctrine, Annex 3-22, is substantively more divergent than aligned with joint doctrine, its tone is nevertheless highly joint.[104] Although it is not described this way, the document reads like a supplement to the ideas presented in JP 3-22, *Foreign Internal Defense*, rather than a stand-alone document.[105] Its discussion is grounded in the context of FID as a joint and interagency endeavor. While briefly providing the airman's perspective, this section also mentions the utility of other services' perspectives during FID operations. Annex 3-22 provides an Air Force–specific discussion of the role of airpower in FID operations without taking on a service-centric tone.

Annex 3-03, *Counterland Operations*, covers interdiction and close air support. The tone of this annex is also joint. There are several explicitly joint statements, such as: "[W]hether air or surface forces are the decisive element is not what matters. Instead, the proper integration of forces is required for successful joint operations."[106] The annex also describes the unique contributions of airpower without describing it as superior. For example, airpower "has proven invaluable in supporting friendly surface maneuvers by destroying, disrupting, delaying, or diverting an enemy's operational military potential." This same section provides examples of how, in some cases, "counterland operations can provide the sole U.S. effort against the enemy" or "counterland operations can decisively engage adversary fielded forces prior to occupation by friendly ground forces."[107] While emphasizing these air component contributions, Annex 3-03 also recognizes contributions of other components: "[W]hen discussing airpower in counterland operations, it is necessary to recognize the contribution of other components' aviation arms to a

---

[102] Annex 3-05, 2017, p. 2.

[103] Annex 3-17, *Air Mobility Operations*, Washington, D.C.: Headquarters, U.S. Air Force, April 5, 2016, p. 2.

[104] Annex 3-22, 2015.

[105] JP 3-22, *Foreign Internal Defense*, Washington, D.C.: Joint Chiefs of Staff, July 12, 2010. The lead agent for JP 3-22 is USSOCOM.

[106] Annex 3-03, 2017, p. 17.

[107] Annex 3-03, 2017, p. 4.

unified effort. Navy, Marine Corps, Army, and SOF aviation assets can be used for both air interdiction (AI) and close air support (CAS)."[108]

Annex 3-03 also includes the "Integration with Surface Maneuver" section, which explicitly promotes an integrated mindset. This section states that "an important factor in successful [AI] operations is integrating air maneuver with surface maneuver. Planning and conducting AI and surface operations within a coherent framework enhances their synergistic effect in those operations involving air and surface forces."[109]

Some Air Force Doctrine Is More Joint Than Service-Centric in Tone

Other Air Force doctrine documents provide some joint context and statements, but they have service-centric elements that prevent them from being considered joint in tone overall. Annex 2-0, *Global Integrated Intelligence, Surveillance & Reconnaissance Operations*, explains that the combatant commander coordinates all assets in theater. It also describes how the JFC establishes execution priorities for the COMAFFOR.[110] Annex 2-0 describes the JIPOE process as "valuable."[111] Moreover, Annex 2-0 introduces some of the unique attributes of airpower without describing the Air Force as superior. For example, "[A] unique advantage is that several platforms used for collection provide an opportunity to minimize the U.S. footprint. Global integrated ISR assets can be based outside of the [area of interest (AOI)] or sequestered on airfields within the AOI that are relatively isolated from the population."[112] Elsewhere, the annex explains, "[O]ne of the most valuable attributes of airpower is its flexibility, the inherent ability to project power dynamically across the entire operational area."[113] However, at one point, the annex does begin to imply superiority, arguing that flexibility and range allow the Air Force to provide operational and strategic-level intelligence support to the JTF, while other services focus their ISR capabilities only to support tactical operations.[114]

Annex 3-34, *Engineer Operations*, mentions the joint context and JFC's intent frequently. The annex also notes that "engineering operations are increasingly being conducted in joint, interagency, and multinational environments."[115] The "Airman's Perspective" section also

---

[108] Annex 3-03, 2017, p. 16.

[109] Annex 3-03, 2017, p. 33.

[110] Annex 2-0, 2015, p. 18.

[111] Annex 2-0, 2015, p. 14.

[112] Annex 2-0, 2015, p. 46.

[113] Annex 2-0, 2015, p. 31.

[114] Although the Air Force may provide operational and strategic intelligence in many cases, other services may do so as well. Therefore, this statement implies limitations of other services that may not be accurate; Annex 2-0, 2015, pp. 4–8.

[115] Annex 3-34, 2014, p. 2.

explains parallels between the work engineers do in different services. The section does, however, take on a service-centric tone when it explains the superiority of the Air Force system of keeping engineers in the active component and notes the downsides of the Army's and Navy's use of the reserve component for combat support roles.[116]

Air Force public affairs doctrine describes the role of Air Force public affairs in supporting national and joint objectives.[117] Because of the nature of public communication, it also emphasizes that coordination is crucial so as to maintain a consistent message throughout the U.S. government and armed forces. Therefore, Air Force doctrine identifies integration and cooperation with joint forces, other services, countries, and external organizations as a key part of public affairs.[118]

There are, however, some exceptions to the document's joint tone. Annex 3-61 makes the following statement under the "Commander's Responsibility" section that seems put the Air Force in competition with other services and departments within the government:

> As spokespersons for the Air Force, the Department of Defense (DOD), and the US Government, commanders and their representatives play a vital role in building public support for military operations and communicating US resolve to international audiences. Providing the maximum disclosure of timely and accurate information as rapidly as possible enables the commander to seize the information initiative. **Involved and sustained public engagement establishes Air Force information dominance, making the Air Force a preferred source of information to internal and external audiences.**[119]

Annex 3-60, *Targeting*, is also more joint than service-centric in tone. The annex situates air processes in a joint setting and states that the purpose of the AOC is to support the JFC's intent.[120] The annex explains that the air component is responsible for "servicing approved targets regardless of which service or functional component nominates them."[121] However, the document also ignores other service considerations when it advocates for the JFC to delegate certain targeting responsibilities, such as approval of the joint integrated prioritized target list to the air component commander rather than to the joint targeting coordination board, which includes representatives from each component.[122]

---

[116] Annex 3-34, 2014, pp. 3–4.

[117] Annex 3-61, *Public Affairs Operations*, Washington, D.C.: Headquarters, U.S. Air Force, June 19, 2014, pp. 5, 25, 28, 37, 39–42.

[118] Annex 3-61, 2014, pp. 5, 16, 37, 42–44.

[119] Annex 3-61, 2014, p. 3. Emphasis in the original.

[120] Annex 3-60, 2017, p. 18.

[121] Annex 3-60, 2017, p. 17.

[122] Annex 3-60, 2017, p. 58.

Other Doctrine Is More Service-Centric Than Joint

Some Air Force doctrine reflects a mindset that is more service-centric than joint. Ten doctrine documents were more insular (focusing primarily on service-specific issues with little discussion of the role the Air Force would play as part of a joint team) or were disparaging of the other services. Air Force basic doctrine, leadership, EW, and cyber operation annexes provide clear examples of service-centric tone.

Core Doctrine, Vol. I, *Basic Doctrine*, provides several instances where the limitations and weaknesses of the other military services are emphasized but those of air forces are not mentioned. For example, *Basic Doctrine* points out that a "surface-centric strategy" focused on "destruction of hostile land forces and the occupation of territory . . . may not achieve the desired strategic outcome" and "territorial occupation, with its attendant large cultural footprint, may not be feasible or politically acceptable."[123] Sea power's constrained ability "to respond rapidly from one theater to another" and its vulnerability in littoral areas are called out as often creating "political risks [that] outweigh the actual military risks."[124] There is no discussion of any limitations associated with airpower. Instead, it notes that airpower has a "degree of versatility not found in any other force,"[125] as well as an ability to create effects "that are ends unto themselves, not just in support of predominantly land or maritime force activities"[126] and to "influence strategic political outcomes."[127] Noting airpower's uses and strengths in doctrine is not inappropriate, but a more balanced presentation of strengths and limitations of air and other military forces could create a more joint tone.

Core Doctrine, Vol. II, *Leadership*, covers the role of airmen and leadership of airmen. Given the document's focus, it is both appropriate and unsurprising that it explains the unique perspective that airmen bring to the joint setting. However, the document is inconsistent in its tone. In the following passage, for example, the document neutrally describes what makes airmen unique: [128]

> Airmen recognize and value airpower in its application, which is fundamentally different and more flexible than other forms of military power and instruments of national power. . . . Air Force forces are employed at different speeds and closure rates and over greater distances and should be applied by those who appreciate the breadth, scope, and uniqueness of that power across the range of military

---

[123] LeMay Center, 2015a, pp. 30–31.

[124] LeMay Center, 2015a, p. 31.

[125] LeMay Center, 2015a, 2015, p. 31.

[126] LeMay Center, 2015a, p. 26.

[127] LeMay Center, 2015a, p. 31.

[128] The Air Force capitalizes *Airmen* when referring to its own personnel and uses *airmen* for personnel for those in the other services and nations. We do not use that convention in this document; LeMay Center, August 8, 2015b, p. 3.

> operations. Due to the distinctive nature of the capabilities brought to the fight, Airmen see their Service as unique.[129]

Leadership also highlights that the Air Force operates as part of a joint force and quotes Gen Henry H. "Hap" Arnold as stating that the "greatest lesson of this war has been the extent to which air, land, and sea operations can and must be coordinated by joint planning and unified command."[130]

However, in other places, *Leadership* disparages other services or implies Air Force superiority. Elsewhere, the doctrine argues that "air, space, and cyberspace are flexible and dynamic domains that present opportunities and vulnerabilities. . . . The Air Force is the premier American military force capable of overcoming those vulnerabilities in defense of the United States."[131] This statement describes the Air Force as superior in two domains—air and cyberspace—in which other services also operate. *Leadership* implies that whereas the other services tend to think narrowly and tactically, airmen have "more inclusive and comprehensive perspectives that favor holistic solutions over tactical ones."[132] As a result, the overall tone of *Leadership* is more service-centric than joint.

Another example of doctrine that is more service-centric than joint is Annex 3-51, *Electronic Warfare.* Four factors contribute to the annex's service-centric tone. First, the role of EW in Air Force operations is mentioned often, but its role in joint operations is hardly mentioned. For example, the annex's introduction states that all three EW divisions (electronic attack, electronic protection, and EW support) "contribute to the success of air, space, and cyberspace operations."[133] Although this is true, the characterization is incomplete and can leave the reader with the idea that EW is not relevant in land or maritime operations. Second, the annex describes Air Force EW organizations in detail but has very little discussion of joint command or EW organizations that a COMAFFOR staff would encounter in a joint setting. Third, such Air Force–specific terms as COMAFFOR and AOC are sometimes intentionally used instead of joint terms, such as JFACC and *joint air operations center* (JAOC). Finally, examples of the capabilities and potential uses of EW are provided in various places in the annex. The vast majority of these examples are Air Force–centric—for instance, focusing on the effects of EW on enemy radar returns or the ability to find surface-to-air threats for other services to attack. Air Force–specific examples certainly have a place in service doctrine. However, replacing some of these with examples demonstrating Air Force EW being used to facilitate operations in the land or maritime domains would improve airmen's understanding of the joint environment. The combination of these four issues could leave the reader with the mistaken impression that EW is a capability

---

[129] LeMay Center, 2015b, p. 3.

[130] LeMay Center, 2015b, p. 41.

[131] LeMay Center, 2015b, p. 5.

[132] LeMay Center, 2015b, p. 5.

[133] Annex 3-51, 2014, p. 2.

focused on enabling air operations and has little relevance to other components and limited applicability in joint operations.

Annex 3-12, *Cyberspace Operations*, also has a more service-centric tone and focus. The airman's perspective section in Annex 3-12 begins by asserting Air Force superiority in the cyber domain. It claims that airmen take a broader view of war and have a historical linkage to cyberspace:

> Airmen conduct a greater percentage of operations not just over the horizon but globally, expanding operations first through space and now also in cyberspace. Just as air operations grew from its initial use as an adjunct to surface operations, space and cyberspace have likewise grown from their original manifestations as supporting capabilities into warfighting arenas in their own right.[134]

This section then asserts, "[C]yberspace operations should be tightly integrated with capabilities of the air and space domains into a cohesive whole, commanded by an Airman who takes a broader view of war, unconstrained by geographic boundaries."[135]

Annex 3-12 does note the other participants in cyberspace operations: "As a part of a larger networked team, the Air Force should plan and execute in complete concert with other Services, nations, and agencies."[136] However, the annex focuses exclusively on the consideration of how cyberspace operations affect Air Force operations. For example, the document refers only to service considerations when it says that "there is a requirement to balance defensive cyberspace actions within cyberspace with their impact on ongoing air, space, and cyberspace operations."[137] The discussion also focuses on service C2 rather than joint C2.[138]

Air Force doctrine sometimes uses historical vignettes to illustrate a particular point about doctrine or an operational challenge. Unfortunately, some of these vignettes have a service-centric tone. For example, a historical vignette in Annex 3-03, *Counterland Operations*, criticizes Army processes during Operation Iraqi Freedom and argues that they made airpower less efficient. This vignette concerns long-running disagreements about fire support coordination measures (FSCMs), especially the placement of the fire support coordination line (FSCL), the line beyond which air forces could fire without extensive coordination with ground forces. Prior to Operation Iraqi Freedom, the Air Force had developed a new concept for operating short of the FSCL. However, it had not taken steps to coordinate with and win the support of the Army prior to the war. Army forces did not feel comfortable with the new process and continued to

---

[134] Annex 3-12, 2011, p. 17.

[135] Annex 3-12, 2011, p. 17.

[136] Annex 3-12, 2011, p. 38.

[137] Annex 3-12, 2011, p. 9.

[138] Annex 3-12, 2011, p. 34.

demand more-burdensome coordination procedures, in part, to prevent fratricide.[139] In spite of this more complex background, the Air Force vignette only highlighted Army shortcomings:

> Because of the Army's extensive process required for changing linear FSCMs, moving the FSCL proved to be a time-consuming process. Therefore, the initial FSCL was placed well beyond the range of land fires in order to accommodate the anticipated rapid movement of land forces into Iraq. The deep placement of the FSCL hampered the efficiency of airpower. . . . The FSCL should be near the maximum operating range of organic tube artillery since beyond that point air and space power provides the preponderance of effects.[140]

This vignette is service-centric because of its criticism of another service without acknowledging the Air Force's limitations or responsibility.

Some Air Force senior leaders' quotes also contribute to a service-centric tone. In *Core Doctrine,* Vol. II, *Leadership*, a quote from Mitchell, considered the father of the Air Force, pointedly indicates the implied superiority of airmen: "In the development of airpower, one has to look ahead and not backward, and figure out what is going to happen, not too much what has happened. That is why the older services have been psychologically unfit to develop this new arm to the fullest extent practicable with the methods and means at hand."[141]

Summary

Air Force doctrine provides insights into the service's current culture and practices. Our doctrine comparison indicates that, beginning at the *Core Doctrine* level, there is inconsistent Air Force acceptance of joint concepts and a joint mindset. Some Air Force doctrine documents are significantly aligned with joint doctrine and are joint in tone, but a larger number diverge from joint doctrine and have elements that engender a service-centric tone.[142] Many Air Force doctrine annexes are not written in a way that encourages a full understanding and appreciation of how the Air Force operates as part of a joint team.

Furthermore, much of Air Force doctrine lacks sufficient references to, alignment with, and application of joint terms and processes. Differences and gaps in operational planning may be

---

[139] Benjamin S. Lambeth, *The Unseen War: Allied Air Power and the Takedown of Saddam Hussein*, Annapolis, Md.: Naval Institute Press, 2013, pp. 255–260.

[140] Annex 3-03, 2017, p. 85.

[141] LeMay Center, 2015b, p. 39.

[142] Although our scales were designed to measure different characteristics of doctrine documents, an argument could be made that there should be a relationship between a document's tone and alignment. After all, doctrine developers' attempts to be consistent with and draw linkages to joint doctrine could reflect a joint mindset that would also come through in the document's tone. In practice, we found that there is some relationship. Documents with a joint tone, for example, tended to be more aligned with joint doctrine than to diverge from it, and documents that are more divergent from joint doctrine are likely to have a more service-centric tone. However, the two variables are not perfectly correlated. As will be discussed in a later chapter, the remedies for increasing substantive alignment and joint tone can be very different.

most consequential given the importance of planning in joint leadership positions. Therefore, doctrine may currently provide inadequate exposure to the basic joint terms and processes, which makes it more difficult for airmen to operate in a joint environment and further hampers the development of an integrated, joint mindset. As the demands of joint integration increase, these deficiencies may become even more problematic.

3. Current Challenges to Promoting Joint Warfighting Proficiency

The analysis of Air Force doctrine in Chapter 2 suggests that the Air Force has only partially adopted joint constructs and a joint mindset. In Chapter 4, we present detailed recommendations for achieving greater alignment of Air Force and joint doctrine. However, the efficacy of these recommendations depends on how doctrine is valued by the Air Force and how airmen develop joint proficiency today. Therefore, this chapter provides an initial exploratory analysis of two questions. First, to what extent would greater alignment of Air Force and joint doctrine help airmen gain greater warfighting proficiency? Second, how does the way that airmen are formally exposed to or use doctrine affect their level of joint warfighting proficiency? The analysis presented here offers an initial set of observations about the broader context that Air Force leaders need to consider as they identify a holistic approach to strengthening joint leaders and teams. Many of the topics discussed in this chapter merit deeper analysis that is beyond the scope of this study.

Overall, the analysis in this chapter suggests that doctrinal changes, on their own, would only have a limited effect on joint proficiency. Doctrine is not central to Air Force culture, so changes to doctrine may not immediately or significantly change practice or mindsets. Moreover, many other factors currently affect the level of joint proficiency in the force. Many airmen do not have opportunities to regularly use joint constructs or engage deeply with joint doctrine until later in their careers. This is because many airmen spend years becoming proficient in their highly technical occupational specialties, and joint constructs do not appear to be widely used in the documents or planning processes they master during this phase of their careers. Although airmen have opportunities for formal exposure to joint doctrine through education, they do not generally have joint experiences until later in their careers, making it difficult to reinforce the doctrine they learn formally with practice. Insufficient familiarity with joint doctrine and settings could make it even more difficult to achieve the level of joint integration envisioned by joint and Air Force concepts. As a result, policies to address some of these aspects of the development of Air Force officers would likely need to change in tandem with doctrine revisions to promote joint warfighting proficiency.

Challenges to Using Doctrine to Promote Joint Warfighting Proficiency

Although doctrine is usually considered a reflection of current culture and practice, other services have used major doctrine revisions as part of a broader campaign to significantly change their services' practices.[1] Doctrinal changes, however, will only have a significant effect if doctrine is valued in the Air Force today. To assess the value placed on Air Force doctrine, we reviewed its place in Air Force history, analyzed trends in the number of personnel directly involved in doctrine development, and reviewed discussions of doctrine in the Air Force's primary professional journal. We also interviewed those directly involved in developing Air Force doctrine today, including the staff of the LeMay Center and doctrine contacts at all but two of the MAJCOMs.[2] One of those two MAJCOMs, Air Combat Command (ACC), no longer has a doctrine office, but we did speak with those familiar with the decision to eliminate the doctrine staff and ACC's subsequent experience.

This diverse analysis suggests that, under current conditions, the effects of doctrinal changes on joint proficiency would likely be limited because of three factors. First, the literature on military organizational change suggests that, in general, doctrine can reinforce, but not lead, broader change in culture and practice. Other steps, such as senior leadership support and changes to promotion incentives, are likely essential conditions to significant improvements in joint warfighting proficiency. Second, we found that doctrine is not currently central to Air Force culture, so the effects of doctrinal change may not propagate rapidly or substantially. Third, strengthening joint leaders and teams is not currently a primary goal of Air Force doctrine. Elevating this goal could create trade-offs with other aims of Air Force doctrine, such as standardizing terminology across the Air Force.

Literature Suggests That Doctrine Can Reinforce, but Not Lead, Change

Promoting higher levels of joint proficiency could require significant changes to Air Force culture and practices. Extensive and varied past research has shown that military organizations have entrenched processes that make significant changes difficult to implement. Moreover, a service's formative experiences often define its culture and, in turn, deeply affect whether a service or community uses new doctrine in practice.[3] In the specific case of the Air Force, the long fight involved in becoming an independent service has made it especially sensitive to

---

[1] For an example of using a significant doctrinal revision facilitate broader changes in the Army, see John L. Romjue, *American Army Doctrine for the Post-Cold War*, TRADOC Historical Monograph Series, Fort Monroe, Va.: Military History Office, U.S. Army Training and Doctrine Command, 1997, pp. 3, 35, 41–50.

[2] We had several discussions with the LeMay Center staff from July to November 2017. These discussions included in-person and phone interviews, as well as email exchanges. We spoke with MAJCOM doctrine points of contact between September and November 2017.

[3] Posen, 1984; Jack Snyder, *The Ideology of the Offensive: Military Decision Making and the Disasters of 1914*, Ithaca, N.Y.: Cornell University Press, 1989; Deborah D. Avant, *Political Institutions and Military Change: Lessons from Peripheral Wars*, Ithaca, N.Y.: Cornell University Press, 1994; Johnston, 2000; Long, 2016; Adam Grissom, "The Future of Military Innovation Studies," *Journal of Strategic Studies*, Vol. 29, No. 5, 2006.

threats about its autonomy and how airpower is applied.[4] The Air Force also historically struggled to integrate its functional communities and build common purpose within the service, suggesting that promoting joint integration will also be a challenge.[5] Together, this literature suggests that promoting joint proficiency will require more than just aligning Air Force doctrine with joint doctrine.

Because militaries are hierarchical, change usually requires the leadership of those who wield formal authority. In order to elicit broad change, officers who support the change need to be promoted to positions of influence, and promotion criteria need to reward those who adopt the new approach.[6] The literature on organizational change calls such leaders "change agents" and notes their importance in advocating and protecting change.[7] Past research has also shown that military organizations have successfully undertaken such changes when some external factor, such as an increase in an enemy's military capabilities, affects the organizations' ability to achieve their primary mission—winning wars.[8] This suggests that a clear linkage between joint proficiency and Air Force success in a future conflict could help to gain internal Air Force support for change.

Updating doctrine may not, on its own, lead to changes in culture and practice. However, it can reinforce other policy initiatives when doctrine is used to shape education and training. Moreover, to the extent that these other policies lead to acceptance of the new approach, revised doctrine is more likely to be used in the conduct of military operations.[9] Therefore, past research suggests that substantial organizational change is difficult and that doctrine can reinforce, but not replace, other elements of a broader approach.

Doctrine Is Not Central to Air Force Culture

The Air Force faces particular challenges in using doctrine to promote joint constructs and a joint mindset. Our analysis shows that the Air Force has not historically been a doctrine-driven force. Moreover, after an increase in resources in the 1990s, there have been declining resources

---

[4] Carl H. Builder, *The Masks of War: American Military Styles in Strategy and Analysis*, Baltimore, Md.: Johns Hopkins University Press, 1989.

[5] Carl H. Builder, *The Icarus Syndrome: The Role of Air Power Theory in the Evolution and Fate of the U.S. Air Force*, New Brunswick, Conn.: Transaction Publishers, 1994.

[6] Stephen Peter Rosen, *Winning the Next War: Innovation and the Modern Military*, Ithaca, N.Y.: Cornell University Press, 1994, pp. 19–21; Builder, 1989.

[7] W. Henry Lambright, "Leadership and Change at NASA: Sean O'Keefe as Administrator," *Public Administration Review*, Vol. 68, No. 2, March–April, 2008; Sergio Fernandez and Hal G. Rainey, "Managing Successful Organizational Change in the Public Sector," *Public Administration Review*, Vol. 66, No. 2, March, 2006.

[8] Kimberly Marten Zisk, *Engaging the Enemy: Organization Theory and Soviet Military Innovation, 1955–1991*, Princeton, N.J.: Princeton University Press, 1993; Rosen, 1994.

[9] Harald Høiback, "What Is Doctrine?" *Journal of Strategic Studies*, Vol. 34, No. 6, December 1, 2011, p. 892; Olof Kronvall and Magnus Petersson, "Doctrine and Defence Transformation in Norway and Sweden," *Journal of Strategic Studies*, Vol. 39, No. 2, February 23, 2016.

devoted to doctrine development and little discussion of doctrine within the service's top professional journal, *Air & Space Power Journal* (*ASPJ*). Our interviews with current Air Force officers with joint experience, discussed in more detail below, reinforced this analysis. Many airmen noted that their counterparts in other services had much greater exposure to and understanding of doctrine at all phases of their careers. Because doctrine is not central to Air Force culture, changes to doctrine may not propagate quickly within the service or lead to significant cultural changes.

Commitment to Air Force Operational Doctrine Was Slow to Emerge

Airmen began grappling with basic questions about how to fight in the air domain during the early years of the Army Air Corps. During the 1920s and 1930s, several individuals within the Air Corps Tactical School began to develop different aspects of the airpower theory, a set of fundamental principles that would guide the employment of airpower. These ideas, which did not yet address the more detailed topics of operational doctrine, were assembled to create informal basic doctrine and justification for an independent air service.[10]

From that time until 1993, the Air Force focused primarily on basic doctrine that described the Air Force's core principles rather than detailed operational doctrine.[11] Senior leaders did not believe that it was possible to maintain accurate and up-to-date operational doctrine because of the ever-developing nature of airpower and related technologies. The first official operational doctrine was published in September 1953 and not updated until 1965. This doctrine remained in place for decades. Responsibilities for doctrine development also moved frequently within the Air Force—therefore, the service did not gain institutional or individual expertise on doctrine development.[12] Moreover, doctrine often reflected the viewpoints not only of the office or division in which it was held but also of the senior officer in charge at the time. Based on this history, several analysts writing in the 1980s and 1990s argued that operational doctrine was not an integral part of the service's organizational culture.[13]

---

[10] Robert T. Finney, *History of the Air Corps Tactical School, 1920–1940*, Maxwell Air Force Base, Ala.: Research Studies Institute, U.S. Air Force, Historical Division, Air University, 1955, pp. 55–59; Thomas H. Green, *The Development of Air Doctrine in the Army Air Arm, 1917–1941*, Washington, D.C.: Office of Air Force History, 1985, pp. 47–48; James A. Mowbray, "Air Force Doctrine Problems, 1926–Present," *Airpower Journal*, Winter 1995, pp. 2–5.

[11] For more background on early doctrine development within the Air Force, see Robert Frank Futrell, "Chapter 7: The Air Force Writes Its Doctrine, 1947–55," in *Ideas, Concepts, Doctrine: Basic Thinking in the United States Air Force, 1907–1960*, Vol. I, Maxwell Air Force Base, Ala.: Air University Press, 1989b.

[12] For more on early Air Force reorganizations related to doctrine development, see Robert Frank Futrell, "Chapter 3: The Air Force in a Changing Defense Environment," *Ideas, Concepts, Doctrine: Basic Thinking in the United States Air Force, 1961–1984*, Vol. II, Maxwell Air Force Base, Ala.: Air University Press, 1989a.

[13] Air Force Doctrine Center, *A Short History of the Air Force Doctrine Center*, Maxwell Air Force Base, Ala., 2005, pp. 4–5; Bruce L. Curry, *Turn Points in the Air: Using Historical Examples to Illustrate USAF Doctrine*, Maxwell Air Force Base, Ala.: Air War College, Air University, 1997, pp. 2–6, 8–9; Dennis M. Drew, "Inventing a

In 1993, the Air Force created a doctrine center responsible for developing basic and operational-level doctrine, as well as coordinating contributions to joint doctrine. The reasons for the center's creation appear to have been twofold. First, articles in professional journals in this period argued that such a center was needed because Air Force doctrine was underdeveloped. Second, the history of the doctrine center, which later became the LeMay Center, further notes that a desire to influence emerging joint doctrine was also a significant reason for consolidating doctrine development.[14] When the center made the transition from a field operating agency to a direct reporting unit under the CSAF in 1997, then-CSAF Gen Ronald Fogelman stated four main reasons for the new designation and reorganization. First, Air Force doctrine had never "caught up" with the notion that airpower changed the fundamental nature of war. Second, the Air Force continually shuffled doctrinal responsibility between three different organizations. Third, the Air Force did not dedicate the same level of organizational commitment to doctrine as other services, particularly the Army and Marine Corps, resulting in comparatively less institutional knowledge of doctrine. Finally, other services played greater roles in joint doctrine development because of Air Force doctrine's "fragmented" nature.[15] Although the creation of the doctrine center reflected an increased emphasis on doctrine in the 1990s, as the next two sections detail, the commitment appears to have waned.

Resources for Doctrine Development Have Been Declining

Available data and interviews suggest that the number of dedicated doctrine development personnel has been declining. The LeMay Center staff reported that staffing levels have declined notably since the center's creation. We were unable to obtain complete personnel data for the period from the center's founding to today.[16] However, the available data, shown in Figure 3.1, substantiate the recollections of long-serving members of the LeMay Center staff. The figure shows that the total number of personnel working at the doctrine center declined after a peak in the 1990s. During this period, the number of officers declined, but some of their numbers have been replenished with more civilian hires.[17]

Doctrine Process," *Airpower Journal*, Vol. 9, No. 4, Winter 1995; Robert C. Ehrhart, "Some Thoughts on Air Force Doctrine," *Air University Review*, March–April 1980; Phillip S. Meilinger, "The Problem with Our Airpower Doctrine," *Airpower Journal*, Spring 1992; Mowbray, 1995, pp. 2–3, 5, 7–10, 12.

[14] Headquarters, U.S. Air Force, *Point Paper on the Functions of New Air Force Doctrine Center*, Maxwell Air Force Base, Ala., 1996, p. 1; Air Force Doctrine Center, 2005, p. 3; LeMay Center, 2017c.

[15] Air Force Doctrine Center, 2005, pp. 4–5.

[16] Beginning in fiscal year 2007, the doctrine center began reporting its personnel data under Air Education and Training Command; Assistant Secretary of the Air Force (Financial Management and Comptroller), *United States Air Force Statistical Digest, Fiscal Year 2010*, Washington, D.C.: U.S. Air Force, 2011.

[17] Assistant Secretary of the Air Force (Financial Management and Comptroller), 2011; Assistant Secretary of the Air Force (Financial Management and Comptroller), *United States Air Force Statistical Digest, Fiscal Year 2004*, Washington, D.C.: U.S. Air Force, 2005; Assistant Secretary of the Air Force (Financial Management and Comptroller), *United States Air Force Statistical Digest, Fiscal Year 2000*, Washington, D.C.: U.S. Air Force, 2001;

Figure 3.1. Personnel for the Air Force Doctrine Center

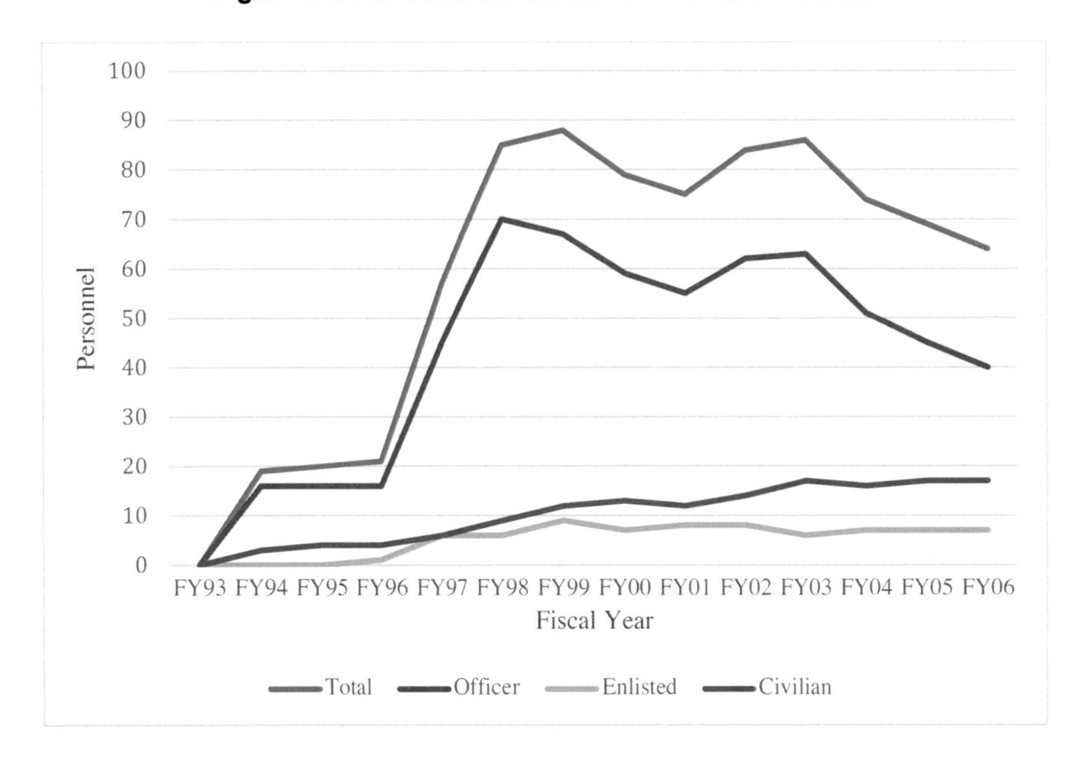

SOURCES: Assistant Secretary of the Air Force (Financial Management and Comptroller), 1997, p. 40; Assistant Secretary of the Air Force (Financial Management and Comptroller), 2001, p. 40; Assistant Secretary of the Air Force (Financial Management and Comptroller), 2005; Assistant Secretary of the Air Force (Financial Management and Comptroller, 2011, p. 40.

The LeMay Center reported that staffing shortages are an ongoing concern. As of July 2017, the LeMay Center reported that it had 23 personnel involved in service, joint, and multinational doctrine development in its doctrine directorate. All 11 civilian positions were filled at that time, but only 11 of 18 authorized active-duty military positions were filled—a 61-percent fill rate. The LeMay Center reported that it struggles to manage its workload in light of personnel shortages.

The number of personnel dedicated to doctrine development at the MAJCOMs has also reportedly declined. In the past, some of the MAJCOMs had more than one person dedicated to doctrine development. However, in recent years, doctrine positions at the MAJCOMs have been cut or discussed as possible areas for personnel cuts when staffs have been downsized. Most notably, ACC is no longer formally participating in the doctrine development process.[18] Most

Assistant Secretary of the Air Force (Financial Management and Comptroller), *United States Air Force Statistical Digest, Fiscal Year 1996*, Washington, D.C.: U.S. Air Force, June 1997.

[18] Interviews with those familiar with ACC's decision to close its doctrine shop reported that the decision was motivated by manning shortages and a perception that the LeMay Center should have sufficient expertise to draw on internally and among Air University students and faculty. Interviews with the LeMay Center staff and doctrine points of contact at MAJCOMs confirm that all other MAJCOMs are still formally involved in doctrine development.

other MAJCOMs have one part-time or full-time person coordinating their MAJCOMs' inputs to doctrine.[19] As with the LeMay Center, many MAJCOM doctrine contacts noted that officer representation in these positions has also declined over time. As a result, the Air Force doctrine development community as a whole has fewer individuals to contribute insights from recent operational experiences. As discussed in Chapter 1, many practitioners at MAJCOMs also contribute to the substance of doctrine. There is no available data on number of people or amount of time devoted to these contributions over time. However, doctrine points of contact at several of the MAJCOMs reported that contributions from these subject-matter experts have also declined.

The declines in staffing for doctrine development should be taken in context: Many other organizations within the Air Force have also faced staffing cuts in this period. Still, doctrine development is among the personnel "bill payers," rather than a priority area that is protected when resources decline. Moreover, although current Air Force officers contribute to doctrine development through most MAJCOM staffing processes, there are few personnel with ongoing and recent operational experience who are dedicated to doctrine development, both at the LeMay Center and at most MAJCOMs. These trends suggest both that doctrine is not currently a service priority and that significant doctrinal changes would likely require additional resources.

Doctrine Is Not a Central Part of the Professional Discourse

To better understand the role of doctrine within the Air Force, we reviewed how often doctrine was discussed in the Air Force's most well-known professional journal, *ASPJ*, and its predecessors.[20] Unfortunately, we do not have data on how widely read the journal is within the Air Force or a way to directly assess whether its articles accurately reflect Air Force priorities. However, since Air Force general officers often take the time to contribute to its pages, the content should be at least a rough indicator of the issues that the Air Force is actively debating.

To assess the prevalence of doctrine discussions, we first reviewed the complete index of articles within *Airpower Journal* (1987–1998), *Aerospace Journal* (1999–2002), and *ASPJ* (2002–fall 2017) and identified articles with the word *doctrine* or *doctrinal* in the title. Figure 3.2 shows the number of these articles over time. We found that 68 articles, or approximately 6 percent of the total number of articles, refer to *doctrine* in the title. The spike in such articles in the mid-2000s reflects a short-lived initiative by *ASPJ* to provide doctrine updates. From fall

---

[19] There are exceptions, such as Air Force Global Strike Command and AFSOC, which have multiple people directly involved in doctrine development.

[20] *ASPJ* is a quarterly peer-reviewed professional journal published by Air University Press and is the self-described "leading forum for airpower thought and dialogue." The journal has changed names several times over the years, including *Airpower Journal* and *Aerospace Power Journal*. We did not include other Air Force publications, such as *Air University Review* or *Air University Quarterly Review*, as part of this analysis; *ASPJ*, "About ASPJ," webpage, undated.

2004 to winter 2007, *ASPJ* included *doctrine NOTAMs* (notices to airmen), which were meant to "succinctly inform Airmen about important changes or concepts and explain their relevance, thus keeping Airmen up to date on our service's constantly evolving doctrine."[21] All but four of the 19 doctrine articles between 2005 and 2007 were NOTAMs or brief updates on specific doctrine.[22] Only one article in the past five years has had *doctrine* in its title.

Figure 3.2. Number of Articles on Doctrine in *ASPJ* over Time

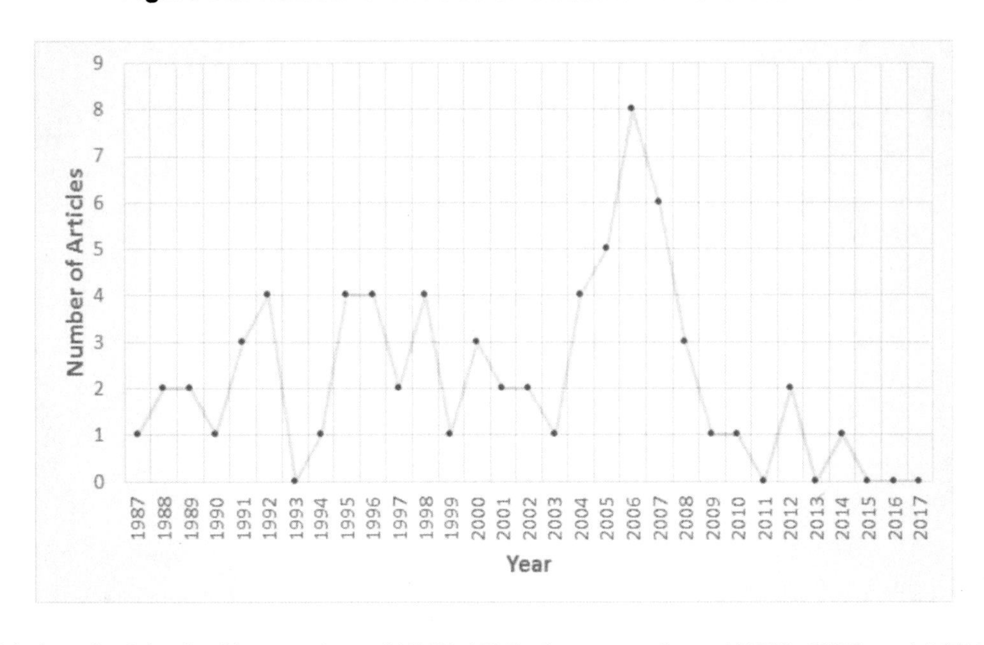

SOURCE: Index of articles for *Airpower Journal* (1987–1998), *Aerospace Journal* (1999–2002), and *ASPJ* (2002–fall 2017).

Second, we searched within the full text of *ASPJ* from September 2002 to fall 2017 and identified 582 articles (or approximately 79 percent of the total number of articles) that include some mention of the word *doctrine*.[23] But on closer examination, most articles reference doctrine in general or the doctrine of other countries (e.g., Russian nuclear doctrine), or cite articles that include *doctrine* in the title. There is less of a focus on the substance of Air Force or joint doctrine in these articles.

Although the discussion of doctrine in Air Force professional journals has been limited in recent years, there were several articles, including two by general officers, on the substance of

---

[21] Paul D. Berg, "Vignettes, Doctrine NOTAMs, and the Latest *Chronicles* Articles," *Air & Space Power Journal*, Vol. 19, No. 1, Spring 2005, p. 17.

[22] Information on recent changes to doctrine along with doctrine advisories and updates can now be found on the LeMay Center doctrine website; see LeMay Center, undated.

[23] This also includes any articles that have *doctrine* in the title. The search only accounts for *ASPJ* articles.

doctrine and doctrine development between 1987 and 1996.[24] The last time a general officer wrote an article about doctrine within *ASPJ* was in 2007, when Maj Gen Charles Dunlap, Jr. (Ret.), penned "Air-Minded Considerations for Joint Counterinsurgency Doctrine."[25]

Overall, there has never been a significant emphasis on doctrine within *ASPJ*. In recent years, the discussion of doctrine appears to have waned even further. The lack of discussion of doctrine in *ASPJ* is another indication that the Air Force is not currently a doctrine-driven service.

Joint Alignment Is Just One of Many Goals of Air Force Doctrine Development

Focusing on the alignment of Air Force and joint doctrine and promoting a joint mindset would be a shift in emphasis from the existing logic of Air Force and joint doctrine development. The LeMay Center described doctrine development as a bottom-up process where each service writes its own doctrine to codify its best practices and war-winning strategies when fighting as a single service. Then, services negotiate on the content of joint doctrine. JPs ultimately reflect the consensus of the services on how they will fight when they combine as a joint force. In general, the LeMay Center takes steps to update Air Force doctrine so that it is consistent with joint doctrine and uses joint terms. However, the LeMay Center does not write service doctrine to be nested under joint doctrine, and joint doctrine does not fundamentally drive the logic or organization of Air Force doctrine.[26] On net, the current doctrine process emphasizes the development of Air Force doctrine first, promoting adoption of these ideas in joint doctrine where possible, and incorporating agreed-on joint constructs into the existing Air Force doctrine framework. However, as the gaps and poor linkages described in Chapter 2 show, joint constructs are inconsistently incorporated for a variety of reasons.

One reason for some of these divergences is that joint alignment is only one of many goals of Air Force doctrine. For example, when the Air Force believes that it has a new best practice that makes sense for the service, that practice is sometimes incorporated into Air Force doctrine. One ongoing example is the current Air Force effort to standardize terminology across Air Force air, space, and cyber communities to make integration across these communities easier. For example, in Air Force and joint doctrine, the term *counterair* refers to "offensive and defensive operations to attain and

---

[24] Richard P. Hallion, "Doctrine, Technology, and Air Warfare: A Late Twentieth-Century Perspective," *Airpower Journal*, Vol. 1, No. 2, Fall 1987; William F. Furr, "Joint Doctrine: Progress, Prospects, and Problems," *Airpower Journal*, Vol. 5, No. 3, Fall 1991; Robert N. Boudreau, "The New AFM 1-1: Shortfall in Doctrine?" *Airpower Journal*, Vol. 6, No. 4, Winter 1992; Drew, 1995; I. B. Holley, Jr., "A Modest Proposal: Making Doctrine More Memorable," *Airpower Journal*, Vol. 9, No. 4, 1995; Ronald R. Fogleman, "Aerospace Doctrine: More Than Just a Theory," *Airpower Journal*, Vol. 10, No. 2, 1996.

[25] Charles J. Dunlap, Jr., "Air-Minded Considerations for Joint Counterinsurgency Doctrine," *Air & Space Power Journal*, Vol. 21, No. 4, Winter 2007.

[26] LeMay Center, 2017c.

maintain a desired control of the air."[27] The LeMay Center now has approval from senior Air Force leaders to use the term *counterspace* instead of the joint term *space control* to describe analogous operations in space.[28] The LeMay Center is planning to integrate this and other Air Force–specific terms into Air Force doctrine. By placing these terms in Air Force doctrine, the LeMay Center expects that the terms will gain credibility to advocate for joint adoption of the new terms. If successful, then alignment between Air Force and joint doctrine would eventually be restored. However, if joint doctrine does not ultimately adopt these terms, Air Force efforts to develop shared terminology internally will create a new source of divergence with joint doctrine.

The current process for Air Force and joint doctrine development emphasizes the development of service best practices but accounts for joint doctrine to the extent possible. The existing process and the priorities within it, therefore, would likely need to change to achieve higher levels of alignment with joint doctrine.

Perceptions of Less Air Force Influence on Joint Doctrine

Many of the people we spoke with throughout this study suggested that the Air Force has not had significant influence on joint doctrine. Some doctrine developers noted that early joint doctrine documents drew heavily on Army doctrine because it was the most well-developed. Although many doctrine developers noted that the Air Force has since made significant contributions, we repeatedly heard comments from others in the Air Force that reflect a belief that the Army has had an outsized role in doctrine development.[29]

A comprehensive assessment of service influence on the substance of joint doctrine is outside the scope of this study. However, on a cursory level, there are elements of doctrine that could be perceived as evidence of disproportionate Army influence. For example, JPs often have callout boxes with quotes from senior U.S. military leaders, politicians, and other military thinkers (see Figure 3.3).[30] Of the 72 quotes from U.S. military leaders contained in unclassified JPs, more than half highlight Army leaders, whereas about 17 percent of the quotes come from Air Force leaders.[31] In reality, the share of quotes from each service likely depends on many factors, not

---

[27] DoD, 2017, p. 55.

[28] LeMay Center, 2017c.

[29] One interviewee captured this sentiment as follows: "How do you spell joint? A-R-M-Y." Other comments noted Army influence by referencing the colors typically associated with the Army (green) and the joint community (purple), such as, "purple is just another shade of green" and "peel back the purple cover on joint doctrine and it is green underneath."

[30] This number only includes the quotes from U.S. military leaders. Other boxes quote foreign political and military leaders, academics, and others.

[31] Of the 72 quotes from U.S. military officers, almost all come from the period since the Air Force became an independent service. Therefore, high Army representation is not a reflection of its longer history. Quotes from individuals who were not U.S. military officers are not included in these data. For those leaders who held senior military positions as well as political positions, such as President Dwight Eisenhower, we only counted quotes made during the years they were senior military leaders.

just service influence. However, such superficial aspects of doctrine may inadvertently contribute to perceptions of Air Force underrepresentation in doctrine.

Figure 3.3. Services of Service Leaders Referenced in Joint Doctrine Callout Boxes

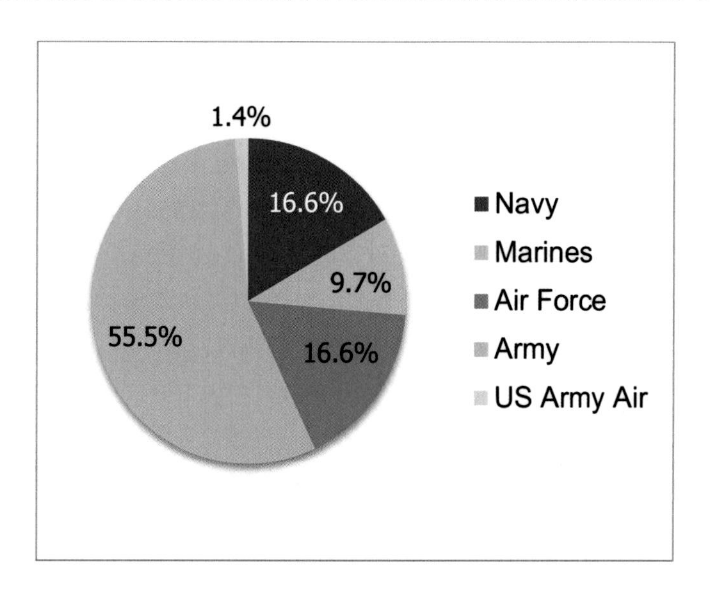

SOURCE: Collected from all joint operational doctrine documents, except for classified or For Official Use Only documents.

Whatever the reality of Air Force influence, perceptions that joint doctrine does not reflect Air Force values or priorities could make it more difficult for the Air Force to build support for aligning Air Force doctrine with joint doctrine. When this view is taken together with evidence presented in the previous sections that doctrine does not have a central role in Air Force culture, our analysis suggests that doctrinal change may be difficult and, on its own, may have a limited impact on the use and acceptance of joint constructs within the Air Force.

Limited Opportunities to Use Joint Doctrine and Constructs

This section asks how airmen learn and use joint doctrine and constructs to identify other potential challenges to promoting joint warfighting proficiency. To understand airmen's formal exposure to doctrine, we reviewed documents on officer professional military education and conducted interviews. Interviewees included staff from Air Education and Training Command (AETC), faculty at Air War College and Air Command and Staff College (ACSC), and students at the School for Advanced Air and Space Studies (SAASS).

To understand the use of doctrine in practice and airmen's overall preparation for joint assignments, we interviewed 24 Air Force officers from a number of communities. During October–November 2017, we spoke in person or by phone with airmen at seven of the ten

MAJCOMs, three of the nine combatant commands, and three of the numbered Air Forces.[32] We also spoke with airmen currently assigned to the Joint Staff and the air component of a geographic combatant command (GCC), as well as an officer involved with AFTTP development. The interviewees included two majors, eleven colonels, six brigadier generals, and five major generals. These airmen had experiences in the combat air force, mobility air force, cyber, nuclear, special operations, and space communities. We focused on active-duty officers, but we also spoke with one from the reserve component.

Our interviewees were not randomly selected. Instead, we used judgment to identify field grade and general officers with appropriate backgrounds from multiple Air Force communities. Although this approach limits the generalizability of some of our findings, it offered a way to learn about officer preparedness across the Air Force within the period of the study. The mix of in-person and phone interviews were semi-structured, meaning that we started with a standard question list but followed up with additional questions and allowed participants to share additional information during the conversation.[33] As discussed earlier, the relatively small number of interviews means that our findings on airmen's preparation for joint assignment are suggestive but not conclusive.

Our analysis suggests that many airmen lack early opportunities to regularly use joint doctrine. Given the highly technical nature of many Air Force positions, many airmen spend most of their early careers becoming technically proficient in their weapon systems. The result is that deep engagement with operational-level doctrine and joint experiences often come much later in their careers compared with other services. Even when airmen were exposed to operational-level doctrinal concepts through education, some airmen we interviewed reported that it was difficult to internalize these concepts when their direct relevance was unclear, and opportunities to apply them were disconnected from airmen's educational experiences.

Airmen Know AFTTP but Have Little Exposure to Operational Doctrine

Airmen generally spend much of their early careers using checklists and tactical procedures to develop platform- or weapon-specific skill sets rather than learning operational doctrine. AFTTP, which provide technical details about many of the Air Force's weapon systems, serve as airmen's basic handbooks.[34] AFTTP documents are inherently tactical and only occasionally

---

[32] These combatant commands were U.S. Cyber Command, U.S. Strategic Command, and U.S. Africa Command. The MAJCOMs we spoke with were ACC, Air Force Space Command, Air Mobility Command, Air Force Global Strike Command, AFSOC, AETC, and Air Force Reserve Command. The numbered Air Forces we spoke with were 8th Air Force, 9th Air Force, and 14th Air Force.

[33] For an overview of semi-structured interviews and judgment sampling approaches, see Margaret C. Harrell and Melissa A. Bradley, *Data Collection Methods: Semi-Structured Interviews and Focus Groups*, Santa Monica, Calif.: RAND Corporation, TR-718-USG, 2009, pp. 27, 32.

[34] Rebecca Grant, "Closing the Doctrine Gap," *AIR FORCE Magazine*, January 1997, p. 51. There are some exceptions to familiarity with AFTTPs.

reach the level of operational doctrine. Still, in theory, these documents could use select joint terms and other constructs to explain the context for tactical missions or to explain how tactical processes may link to joint processes.

We did not assess the level of alignment of AFTTP and joint doctrine, but the LeMay Center reported that there is some divergence. Development of AFTTP is coordinated by such organizations as the 561st Joint Tactics Squadron. AFTTP coordinators rely on tactical experts from the community to write and revise the actual content. The documents are sent to the LeMay Center for input. However, the LeMay Center reported that it does not have sufficient manpower to thoroughly review all AFTTP for alignment with joint or operational service doctrine. For those AFTTP it has reviewed, the LeMay Center has found some inconsistencies with joint doctrine.[35] Because many airmen spend so much time developing tactical skills and using AFTTP with little connection to operational doctrine, airmen may not habitually use joint constructs or doctrine until relatively late in their careers.

Airmen Have Less Experience with Planning Than Their Joint Counterparts

Airmen reported that later exposure to joint planning concepts left them less prepared than their joint counterparts to participate in the joint planning processes. Planning is a critical skill in the current joint environment and may become increasingly important to master early as the joint force moves toward integrating military capabilities at lower levels of command. Several airmen observed that Army and Marine Corps officers are exposed to, learn, and use planning processes similar to the joint planning process from the time they are lieutenants. Additionally, many noted that the Army's planning process, the military decisionmaking process (MDMP), is very similar to the joint planning process. Because soldiers use MDMP throughout their careers, their transition to joint planning is less of a conceptual leap. Meanwhile, airmen use Air Force–specific mission planning processes as they master their weapon systems, and the terminology and steps involved in those processes are different from those in the joint planning process.

Moreover, officers reported that the current manifestations of the Air Force's C2 philosophy, centralized control, decentralized execution, and the ATC consolidate many planning functions in the AOC. Airmen who do not serve in an AOC may not have many opportunities to practice joint planning processes or apply joint planning principles. On net, this suggests that airmen have fewer opportunities to develop skills in joint planning compared with their counterparts from other services. These reports from Air Force officers are consistent with the lower emphasis on planning we found in our detailed review of Air Force doctrine documents.

---

[35] LeMay Center, 2017c.

Doctrine Education and Application Are Not Always Linked

The services are responsible for integrating joint professional military education (JPME) into the curriculum of their professional military education programs, starting with precommissioning education. Detailed joint curriculum requirements for officers begin with JPME Phase I, which is completed around the rank of major. JPME Phase I is integrated into the curriculum at intermediate-level service schools, such as ACSC or the Army's Command and General Staff College. A smaller subset of Air Force officers who are seeking to become joint qualified officers may also complete additional joint education as colonels or lieutenant colonels. JPME Phase II is integrated into the curriculum of the service war colleges and the Joint Advanced Warfighting School (JAWS) and offered through shorter dedicated courses at the Joint and Combined Warfighting School. Capstone courses for general officers are offered through National Defense University.[36]

Although airmen have these educational opportunities, there does not appear to be a purposeful linkage between JPME and joint experience that would help airmen internalize joint doctrine and constructs. By law, service members must have a certain combination of joint education and joint experience to become a joint qualified officer and to be eligible to become a general or flag officer. However, the services are generally responsible for how JPME is sequenced with joint experiences. So, with a few exceptions, there is no legal or policy requirement to have a joint assignment right after JPME or even to have JPME completed prior to a joint assignment.[37]

Air Force policy documents discuss the need to develop airmen's understanding of joint operations through education, but they do not discuss the role of joint experience in developing this competency.[38] We have not been able to obtain data on how JPME and joint experiences are sequenced in the Air Force. However, Air University faculty and Air Force officers we interviewed reported that there is no formal system for linking JPME and joint experience. As a

---

[36] CJCSI 1800.01E, *Officer Professional Military Education Policy*, Washington, D.C.: Joint Chiefs of Staff, May 29, 2015. The LeMay Center did a survey in 2015 to understand how Air Force schools use doctrine. This showed that other schools, such as the Air Force Institute of Technology's School of Systems and Logistics, also use joint doctrine in their curricula. SAASS does not emphasize doctrine, assuming that students get that exposure through earlier educational experiences; LeMay Center, 2017a. Some airmen may also, for example, take training specifically on joint topics that are offered through the LeMay Center; Lemay Center, "Senior Leader Courses," webpage, Air University, April 27, 2017b; Lemay Center, "Intermediate Courses," webpage, Air University, April 27, 2017a. Unlike the other services, the Air Force does not have an advanced planning school. SAASS develops strategists rather than operational planners; Spirtas, Young, and Zimmerman, 2009, pp. 83–84.

[37] As an example of an exception, JPME does have to be completed in advance if a position is listed as critical on the joint duty assignment list. Students must also go to a joint assignment after graduating from JAWS; CJCSI 1800.01E, 2015.

[38] The Air Force has a model for developing the competencies of individual airmen through education, training, and experiences known as the continuum of learning. However, Air Force policy documents only discuss joint competency development through education; Annex 1-1, 2017, p. 5; Air Force Manual 36-2647, *Institutional Competency Development and Management*, Washington, D.C.: Headquarters, U.S. Air Force, March 25, 2014, incorporating Change 2, September 15, 2016, pp. 3, 12, 21–25.

result, airmen do not always get a chance to reinforce the doctrine they learn through education with practical experience. Moreover, several of the officers we interviewed reported that they would have been more prepared for their joint assignments had they completed JPME Phase II prior to (rather than during or after) their joint assignments. Others who did happen to have JPME Phase II complete prior to their first joint assignment noted that it was extremely helpful for understanding joint processes and terms. These interviews provide preliminary evidence that improved linkages between JPME and joint experience could help airmen develop higher levels of joint proficiency.

Joint Experience Often Occurs Later for Airmen

Interviewees repeatedly told us that earlier joint experience is critical—they did not really learn joint doctrine, joint constructs, or a joint way of thinking until they had joint experience. For many airmen, joint experience comes late in their careers, making it difficult to master joint doctrine and constructs before taking a more senior joint assignment. Others cited interactions with other services, even those not formally qualifying as joint assignments, as formative in the way they thought about and approached problems in a joint setting.

One of the assumptions behind the CSAF's initiative on strengthening joint leaders and teams is that airmen have lower levels of joint warfighting proficiency than their counterparts in other services. Therefore, airmen must spend considerable time learning to speak the joint language, understand joint processes, and integrate across domains to "catch up" to their peers from other services. We asked officers currently or previously in joint positions whether they were prepared for joint assignments and how it practically affected their ability to integrate into, influence, and lead joint teams.

Our interviews revealed variations in airmen's preparation for joint assignments. Some airmen reported having their first joint experiences later in their careers—many at the colonel level—and having to learn joint processes and terms on the job. Similarly, an airman currently at a combatant command noted that he has sent airmen to a ten-week JPME Phase II course during their assignments because they did not arrive with an understanding of how to operate in a joint setting. Several interviewees emphasized that it was not just terms and processes but also an understanding of how each service contributes and solves problems that airmen needed to develop to carry out their joint assignments.[39]

Conversely, airmen with earlier exposure to joint settings described very different experiences. Airmen in special operations careers, for example, reported persistent exposure to joint settings through exercises and operations throughout their Air Force careers. Those who had served as air liaison officers (ALOs) reported that they had learned joint processes and how

[39] Relatedly, a separate RAND study that included interviews with current and former leaders from other services and the Office of the Secretary of Defense found that insufficient joint experience affected airmen's selection for senior joint positions (see Lee et al., 2017, pp. 25–27).

other services think and solve problems. These interviewees said that these experiences had made it easier to integrate into joint settings at a more senior level.

Summary

This chapter has used several approaches to understand the potential role of Air Force doctrine in promoting joint proficiency. The relatively short history of operational doctrine within the Air Force, declining personnel levels, and low levels of discourse on doctrine in *ASPJ* all suggest that operational Air Force doctrine is not central to Air Force culture. Because all of these indicators point in the same direction, they strongly suggest that doctrinal changes on their own may only have a limited effect on joint proficiency. The detailed analysis presented in Chapter 2 and discussions with the LeMay Center also suggest that the current Air Force doctrine development process sometimes prioritizes internal Air Force goals and promotion of Air Force ideas within joint settings over joint alignment. As a result, using Air Force doctrine to promote joint proficiency could be challenging, require additional resources and senior-leadership attention, and entail a change in emphasis within the doctrine development process.

We also asked whether there are other aspects of how airmen learn or use doctrine that currently affect the level of joint proficiency within the Air Force. Although this analysis was only exploratory, it raised several issues for the Air Force to consider as part of a holistic approach to strengthening joint leaders and teams. In particular, airmen reported that they do not use joint doctrine and constructs regularly throughout their careers, joint experiences often come late, and JPME is often disconnected from joint experiences. Together, these findings suggest that strengthening joint proficiency will also require creating more opportunities to use joint doctrine and constructs in practice.

4. Recommendations to Promote Joint Warfighting Proficiency

This chapter recommends ways to respond to the CSAF's call for the Air Force to "gain proficiency in joint warfare earlier in the careers of Airmen."[1] This study began with the premise that being able to speak about and understand joint constructs—the principles, processes, and terms found in joint doctrine—and adopt a mindset that the Air Force is part of a joint team are critical foundations for higher levels of joint proficiency. The following recommendations, which are divided into two sets, are intended to help the Air Force strengthen these foundations. The first set of recommendations relates to the substance of Air Force doctrine and the doctrine development process. As discussed earlier, doctrine is not central to Air Force culture, so changes to the content of doctrine may only have limited effects on their own. An initiative to raise the status of doctrine in the Air Force could magnify the impact of doctrinal revisions over time.

The second set of recommendations is focused on how to expose airmen to joint doctrine, constructs, and experiences throughout their careers as part of a broader program of force development. This second set of recommendations should be a higher priority because habitual use of joint constructs would likely have a greater effect on airmen's preparation for joint positions.

These recommendations, if implemented, would require additional resources and may run counter to other Air Force goals. As such, Air Force senior leaders will want to assess how strengthening joint leaders and teams weighs against other initiatives. The chapter concludes with a discussion of the timeline on which these recommendations might be implemented and the relative difficulty of implementing them.

Recommendations for Doctrine Content and Development

As Chapter 2 showed, there are opportunities to increase the alignment of Air Force and joint doctrine. Before doing so, the Air Force needs to decide how it will prioritize increasing joint doctrine alignment with other goals for Air Force doctrine. Once it has made that decision, there are several ways, ranging from minor to more significant, to change the substance and tone of Air Force doctrine. Although these changes would reinforce a shift toward greater joint proficiency, the immediate effect of doctrinal changes might be limited by the current low profile of doctrine within the service. In the longer term, putting more resources toward doctrine development and a greater emphasis on doctrine within the Air Force could lead to more significant effects.

---

[1] Goldfein, 2016.

Decide on the Priority Objectives for Air Force Doctrine

Below, we identify several specific ways to better align Air Force doctrine with joint doctrine. Some of these recommendations, such as using graphics to show the relationships between joint and Air Force processes, would require minimal additional investment (e.g., staff time, funding, or leadership attention). However, other changes, such as reorganizing the internal structure of Air Force doctrine documents to more closely parallel joint doctrine, would require a more fundamental change in approach to doctrine development and potentially involve more trade-offs.

Therefore, Air Force leaders need to weigh the benefits of these changes to strengthening joint leaders and teams against other goals for Air Force doctrine. For example, as discussed in Chapter 3, the LeMay Center is currently developing new terminology for space and cyber operations to align it with existing terminology for the air domain. Although this initiative serves the goal of greater Air Force integration, such differences may mean that airmen have less exposure to relevant joint constructs. A decision by senior leaders about the priorities for Air Force doctrine could help the LeMay Center navigate these trade-offs.

Minimize and Explain Differences Between Air Force and Joint Doctrine

Air Force leaders may decide that certain priorities necessitate differences between Air Force and joint doctrine. The two steps described in the following sections could help to reduce confusion over these differences and improve airmen's preparation for joint settings.

Explain the Nature of and Logic Behind Divergent Air Force and Joint Constructs

There could be cases where Air Force leaders decide that the Air Force needs to use constructs that diverge from joint constructs. For example, the Air Force has long held that the flexibility and scarcity of airpower necessitate a C2 philosophy of centralized control and decentralized execution. However, as discussed in Chapter 2, the resulting differences with the joint philosophy of mission command are not clearly explained. Promoting joint proficiency does not require the Air Force to eliminate all differences with joint doctrine. Rather, in these cases, the Air Force should clearly explain both the nature of the differences and the logic behind the alternative Air Force construct.

Consider Mechanisms Other than Doctrine to Introduce New Constructs If They Are Inconsistent with Joint Doctrine

Air Force doctrine documents are sometimes divergent from joint doctrine because the Air Force has identified what it sees as a new best practice or better terminology. The Air Force can update its own doctrine more rapidly than joint doctrine can be changed. Therefore, the LeMay Center said that Air Force doctrine may be temporarily ahead of and out of sync with joint doctrine. The Air Force generally tries to persuade the joint community to incorporate these practices into joint doctrine. Having a new practice enshrined in service doctrine also reportedly makes these ideas more credible to the joint community. As a result, such differences could be short-lived if the next revision of joint doctrine accepts the change. Alternatively, the Air Force could theoretically revert to the previous version of doctrine if the joint community did not accept the Air Force's proposals. However, as discussed in Chapter 2, there are some divergences between Air Force and joint doctrine that have persisted for a longer period of time. For example, Air Force cyberspace doctrine has several differences from joint doctrine and has not been updated since 2011.

The Air Force can and should continue to innovate to propose new constructs for joint doctrine. It could, however, propose new practices and terminology using Air Force concept documents, rather than by including them in Air Force doctrine. These concept documents could still be promulgated internally so that airmen could be aware of emerging best practices, but they could make it clearer to airmen that these practices have not yet been adopted by the joint community. If these documents had senior Air Force leadership backing, they could be a credible way of advancing new doctrinal ideas within the joint community, without introducing potentially conflicting information into doctrine documents. Based on this approach, the Air Force could socialize conceptual and terminology changes through white papers and test the changes in joint exercises and wargames to gain joint feedback *before* introducing the changes in service doctrine. However, if this or another mechanism cannot carry the same weight as Air Force doctrine in joint doctrine discussions, then this recommendation might create a trade-off: Air Force and joint doctrine would be more aligned, but the Air Force might have somewhat less influence in some doctrinal debates in the joint community.

Revise Air Force Doctrine to Reinforce Connections with Joint Doctrine

As Chapter 2 showed, Air Force doctrine is uneven in its use of joint constructs. This section outlines several options for increasing the alignment of Air Force doctrine with joint doctrine if the Air Force elects to make this a priority. In broad terms, we recommend rethinking the relationship between Air Force and joint doctrine. Currently, Air Force doctrine is often developed to address service concerns and to influence joint doctrine. Joint doctrine does feed into Air Force doctrine, but it does not seem to fundamentally shape its structure or logic. If, as the CSAF stated, increasing the joint warfighting proficiency of airmen is a top priority, then joint doctrine (the best practices for joint warfighting) should be a critical input to Air Force doctrine. The recommendations here focus on specific ways that joint doctrine could be more deeply reflected in Air Force doctrine.[2]

---

[2] Some of these recommendations would require an increase in the resources devoted to doctrine development—for example, additional doctrine developers at the LeMay Center. We return to the question of the priority of doctrine within the Air Force later in this chapter.

Align the Internal Organization of Annexes with Joint Publications

Currently, the internal organization of Air Force and joint doctrine are very different, making it difficult to find related information. Revising the internal organization of Air Force doctrine documents to more closely follow the internal organization of joint doctrine would be one way to make it easier for airmen to find related information. This would include, to the extent possible, aligning the order in which topics are presented.[3]

Moreover, the Air Force should follow joint doctrine in providing clearer connections between doctrine topics. Providing explicit connections and transitions between DTMs would help the reader to understand how they fit together. The Air Force should adopt the joint approach and write annexes as books that have a strong internal logic and clear connections between sections. Furthermore, the Air Force should revise documents holistically, rather than simply focusing on a subset of DTMs. Although select DTMs might be the focus of a particular revision, the process should include a review of the entire annex to ensure that its internal logic is consistent.

Summarize Key Joint Concepts and Reference Joint Publications

Air Force doctrine developers currently seek to minimize repetition in joint doctrine with shorter documents that airmen are arguably more likely to read. However, some limited introduction of joint concepts, processes, and terms is necessary to orient airmen to joint warfighting. Although in-text hyperlinks to JPs are helpful for an airman who wants additional information, these features are not a substitute for brief summaries of joint doctrine and constructs within the Air Force doctrine. An annex should be explicit if an airman needs to read an entire JP or section before reading an annex or DTM. Air Force annexes should provide in-text references to relevant JPs, not just hyperlinks. There are already several noteworthy cases of useful integration of joint concepts in existing Air Force doctrine that could be used as examples, such as the special operations operational annex.

Include More Information on Planning Processes and Considerations

As discussed in Chapter 2, Air Force planning documents tend to have less emphasis on planning than their joint counterparts. Moreover, in our interviews, airmen repeatedly pointed to deficiencies in their preparation to participate in joint planning. Including clear linkages to joint planning doctrine, as well as information on planning considerations in Air Force doctrine, is an important first step toward preparing airmen in this area. Planning is central to the language of joint warfighting. The Air Force should more proactively integrate joint planning terminology and other constructs in its doctrine and, when appropriate, AFTTP to give airmen the basic communication tools as early as possible.

---

[3] Air Force doctrine may not always cover the same topics as joint doctrine, so exact alignment is unlikely.

Include Graphics to Show Connections Between Air Force and Joint Doctrine

Graphics are another way to help airmen understand the relationship between Air Force and joint processes and principles that would not substantially affect the length of the documents. As discussed in Chapter 2, Air Force targeting doctrine has a helpful graphic that shows how the steps of the ATC relate to the JTC (Figure 2.4).[4] Joint doctrine also frequently has diagrams that show the relationship between national- and operational-level organizations and processes that could serve as models for similar comparisons between service and joint doctrine.

Where Relationships or Processes Are the Same, Use Joint Graphics to Reinforce Alignment

In some cases, Air Force annexes describe relationships or processes that are consistent with joint doctrine. In these cases, using the joint diagram, rather than creating a new Air Force diagram, could reinforce the alignment.

Revise Historical Vignettes and Select Senior Leaders' Quotes to Show Air Force Contributions as Part of a Joint Team

Reviewing and revising vignettes with attention to how other services and the joint force are portrayed could improve the overall tone of Air Force annexes. For example, the vignette described in Chapter 2 could be revised to illustrate the importance of decisions about fire support coordination measures while still maintaining a joint tone. The following proposed revision notes the joint team's shared responsibility for failing to find a system that could balance air efficiency with ground force protection:

> The joint force lacked an efficient system for changing FSCMs. The resulting deep placement of the FSCL and the time-consuming coordination procedures created sanctuaries for Iraqi forces to operate. . . . The joint force needs to develop an improved FSCM system that both protects ground forces and efficiently responds to fleeting targets beyond the range of organic tube artillery.

Our suggested revision could add nuance that helps an airmen better understand the concerns of counterparts on a joint team.

As noted in Chapter 2, the Air Force doctrine also has some senior leaders' quotes that note the value of operating in a joint setting. Using more quotes like these and removing senior leaders' quotes that emphasize Air Force superiority at the expense of other services could also help to promote a joint mindset.

Use Outside Reviewers to Assess Content and Tone

After substantial revisions to Air Force doctrine documents, a review by an outside expert or experts could be used to assess both substantive alignment and overall tone. Appropriate reviewers could include faculty at Air University or other experts in air operations who have not been involved in drafting a given doctrine document. Because doctrine development includes

---

[4] Annex 3-60, 2017, p. 55.

input from a large number of people, a fresh set of eyes on a document is especially important to ensure that its overall internal logic is sound. This recommendation would require additional resources for the doctrine development process, but it would also be a clear indicator of senior leadership commitment to the focus area objective.

Elevate the Status of Doctrine Within the Air Force

The analysis in Chapter 3 shows that operational doctrine has not traditionally been an Air Force priority. As a result, using doctrine to promote joint constructs and a joint mindset is unlikely to have significant effects on joint proficiency in the short term. It would likely take a sustained and significant effort to elevate the status of doctrine and the effect of doctrinal changes over the longer term. However, doing so could strengthen joint leaders and teams in several ways.

First, more exposure to doctrine at all levels of JPME could help airmen better articulate Air Force contributions to the joint team. Doctrine is the language of the joint community. The first step to being effective in a joint setting is being an expert in what your service can contribute and how it can be integrated with the contributions of other services. In other words, greater mastery of Air Force doctrine is a first step toward being effective on a joint team.[5]

Second, if the Air Force more deeply engages with its own doctrine, it may also become more influential in shaping joint doctrine. Many airmen pointed out that the Army had historically had significant impact on joint doctrine because the Army already had well-developed service doctrine. If the Air Force put more resources toward doctrine development and awareness, it may gain greater influence through increased doctrinal proficiency and credibility.[6]

Third, dedicating more resources to doctrine development and engaging more people with recent or ongoing operational experience in doctrinal debates could lead to better Air Force doctrine. As one analyst put it, developing and debating doctrine helps "sharpen our language and clarify our concepts."[7] In the recent past, the United States has been engaged in counterinsurgency and steady-state operations that do not replicate traditional full-scale conventional war against a near-peer competitor. One of the values of doctrine is that it provides the military with validated and widely accepted approaches to start with when the next,

---

[5] LeMay Center, 2017a.

[6] Thomas P. Ehrhard, *An Air Force Strategy for the Long Haul*, Washington, D.C.: Center for Strategic and Budgetary Assessments, 2009, p. 57.

[7] Harald Høiback, "The Anatomy of Doctrine and Ways to Keep It Fit," *Journal of Strategic Studies*, Vol. 39, No. 2, 2016, p. 192.

potentially very different, conflict begins. Deeper engagement with doctrine now could give the Air Force a better starting point for the next conflict.[8]

Finally, deeper familiarity with doctrine could also help airmen operate in a conflict in a more contested environment. For example, the *Air Force Future Operating Concept* discussed the possibility that future operating conditions may require less centralized control and greater initiative at lower levels.[9] The common framework that doctrine provides could help facilitate mission-type orders and reduce need for extensive coordination. The common framework, if mature and fully understood by airmen, would allow commanders to focus on communicating the unique aspects of a mission and keep their communications brief.[10] Putting greater resources toward and increasing engagement with doctrine development like this would come at a cost, but doing so could improve Air Force contributions in future operations and its integration into the joint team.

Several steps could help to elevate doctrine within the Air Force and increase the effect of doctrinal changes. Including a discussion of doctrine at a Corona meeting (a gathering of senior Air Force leaders that takes place three times a year) would signal the importance of doctrine. Discussion of doctrine in official statements and at such events as Air Force Association meetings could reinforce this initiative. General-officer engagement on doctrinal questions and articles in professional journals, such as *ASPJ*, could also spark greater engagement with doctrinal debates. Although the LeMay Center commander is a general officer, our interviews suggest that doctrine development staff positions are not perceived as premier, career-enhancing assignments. Changing those perceptions by assigning more top talent with recent operational experience to the LeMay Center's doctrine development division and by promoting more of those who serve there and in other doctrine-related positions would demonstrate greater senior leadership commitment and could increase the credibility of doctrine within the Air Force.[11] More personnel involved in doctrine development—through additional staffing for doctrine development at the LeMay Center or through a higher priority on doctrine contributions from the MAJCOMs—might also be necessary.

---

[8] On the value of doctrine in managing uncertainty and preparing for the next conflict, see Barry R. Posen, "Military Doctrine and the Management of Uncertainty," *Journal of Strategic Studies*, Vol. 39, No. 2, 2016, pp. 165–167; Høiback, 2011, p. 890.

[9] U.S. Air Force, 2015, pp. 10, 12, 21. For a discussion of how conditions affect the appropriate amount of centralized control, see Harvard, 2013.

[10] Posen, 2016, pp. 165–167; Høiback, 2011, p. 890; Headquarters, Department of the Army, 2014, pp. 1–4.

[11] The 2015 Air University *Strategic Plan* made a similar recommendation, noting the need to "recruit and retain high-quality doctrine action officers." Air University, *Strategic Plan*, Maxwell Air Force Base, Ala.: Air University Press, September 2015, pp. 17–18.

Increase Exposure to and Experience with Joint Doctrine and Constructs

As discussed in Chapter 3, most Air Force officers are formally introduced to joint doctrine through professional military education and, in some cases, dedicated training. Although there is room for additional dedicated training, the recommendations here focus on how to enhance joint warfighting proficiency by using joint doctrine more frequently in practice. In particular, we recommend integrating joint terms and other constructs as part of regular Air Force activities and linking joint education with joint experiences.

Integrate Joint Terms and Other Constructs into Regular Air Force Activities

Given that many airmen spend a good deal of time developing their technical proficiency, another way to prepare airmen to speak the joint language is to introduce joint terms and other constructs into these tactical activities so they are used habitually. Two possible avenues would be introducing more joint terms and other constructs into AFTTP and identifying opportunities to use joint planning processes for regular Air Force activities.

AFTTP documents generally cover highly technical topics and have little direct linkage to operational-level doctrine. However, there are places, such as in describing a platform's mission sets, where using joint constructs could be appropriate. Additionally, the 561st Joint Tactics Squadron is currently developing TTP for integrating multiple weapon systems for mission areas, such as suppression of enemy air defenses (SEAD). These documents may offer particularly good opportunities to introduce joint terms and processes or reference relevant JPs. Implementing this recommendation would require additional staffing for the LeMay Center to review AFTTP for consistency with joint doctrine and to identify opportunities to introduce joint terms and other constructs.

A second approach would be to identify opportunities to use joint planning processes for Air Force activities. As discussed in Chapter 3, airmen repeatedly reported that they had little to no experience with joint planning processes prior to joint assignments. Revising service-specific planning processes to use as much of the joint planning terminology and processes as appropriate could help make airmen more conversant once they reach a joint assignment.

Finally, the Air Force could review its tactical exercises for more opportunities to employ joint terms and other constructs.[12] For example, the Air Force could identify ways to use more joint constructs during the Weapons School Integration exercise, the capstone to the weapon school for the Air Force's elite instructors that also serves as the proving ground for advanced TTP. Further emphasizing joint constructs in this high-profile exercise would reinforce the

[12] Some of these exercises already have a joint component. For example, during Green Flag exercises, the Air Force supports Army forces training at the National Training Center (NTC). A review could identify additional ways to integrate joint terms, processes, and principles; Nellis Air Force Base, "Exercises and Flight Operations," webpage, undated.

importance of joint constructs and give the Air Force's weapon systems experts more exposure to the language of the joint force.

Link Formal Exposure to Doctrine with Joint Experiences

Currently, there is no purposeful link between an officer's formal exposure to joint doctrine during JPME and a joint experience that could reinforce these concepts. This could be because JPME and joint experiences are sometimes seen as boxes that airmen need to check to become a general officer rather than desired education in the development of all airmen. Although creating such a linkage might not be possible for all officers, at least some subset of Air Force officers might be able to advance their joint warfare proficiency. Even shorter experiences, such as training with an Army unit during an NTC rotation, could offer opportunities to reinforce the joint doctrine that officers learn with practice and to see how other services approach problems.[13] Another option would be to create brief refresher courses—online or in-residence—covering joint doctrine, terminology, and other constructs. Airmen could increase their familiarity by taking these refresher courses on their way to joint assignments. Although linking JPME with joint assignments would be the preferred approach, these brief refresher courses could be helpful in cases where that linkage is not feasible.

Conclusion

The recommendations we offer above are a mix of potential short-term wins and longer-term structural changes. There are several straightforward ways to quickly increase the alignment of Air Force and joint doctrine. Such steps as providing brief summaries of joint concepts, including references to joint documents, and creating graphics to show connections between Air Force and joint doctrine do not require substantial shifts in resources or current practices. However, because of the relatively low emphasis on service doctrine within the Air Force, these may also have a relatively minimal effect in the short term.

Some of the changes that may have a greater impact on joint warfighting proficiency are also more resource intensive and would involve more-significant commitment to this focus area. Deeper changes to the structure of Air Force doctrine, for example, may require more-difficult decisions about how to prioritize service versus joint integration. Similarly, more purposefully linking joint education and joint experiences could be in tension with the career paths that are traditionally valued in the Air Force and could create challenges in the personnel assignment system. Therefore, implementing these recommendations would require continued senior leadership clarity on how to manage these trade-offs.

---

[13] An initiative under way at AETC to create a range of immersive experiences, known as developmental special experiences, appears to be consistent with this recommendation. Some of these experiences will be with services or in joint settings and will range from one day to one year.

Other recommendations, such as adopting joint planning terminology and processes, could also be a significant cultural shift for many in the Air Force. The Air Force has historically been very protective of its autonomy and unique service culture. Therefore, overcoming this resistance may require sustained attention from senior Air Force leaders and a clear explanation of how improving joint warfighting proficiency will help the Air Force become a more effective fighting force.

The recommendations offered here demonstrate that the Air Force has multiple doctrine-related options that could increase the joint warfighting proficiency of airmen today and into the future as the demands of joint integration increase. Ultimately, the first step for the Air Force is to decide how this focus area fits into the overall priorities of the Air Force.

References

Adamsky, Dima, *The Culture of Military Innovation: The Impact of Cultural Factors on the Revolution in Military Affairs in Russia, the U.S., and Israel*, Stanford, Calif.: Stanford University Press, 2010.

AFI—*See* Air Force Instruction.

Air & Space Power Journal, "About ASPJ," webpage, undated. As of October 31, 2017: http://www.airuniversity.af.mil/ASPJ/About/

Air Force Doctrine Center, *A Short History of the Air Force Doctrine Center*, Maxwell Air Force Base, Ala., 2005.

Air Force Instruction 10-13, *Air Force Doctrine*, Washington, D.C.: Headquarters, U.S. Air Force, August 25, 2008.

Air Force Instruction 10-1301, *Air Force Doctrine Development*, Washington, D.C.: Headquarters, U.S. Air Force, June 14, 2013, incorporating Change 1, April 23, 2014.

Air Force Instruction 10-2801, *Force Development Concepts*, Washington, D.C.: Headquarters, U.S. Air Force, October 23, 2014.

Air Force Manual 36-2647, *Institutional Competency Development and Management*, Washington, D.C.: Headquarters, U.S. Air Force, March 25, 2014, incorporating Change 2, September 15, 2016.

Air University, *Strategic Plan*, Maxwell Air Force Base, Ala.: Air University Press, September 2015.

Annex 1-1, *Force Development*, Washington, D.C.: Headquarters, U.S. Air Force, April 17, 2017.

Annex 2-0, *Global Integrated Intelligence, Surveillance & Reconnaissance Operations*, Washington, D.C.: Headquarters, U.S. Air Force, January 29, 2015.

Annex 3-0, *Operations and Planning*, Washington, D.C.: Headquarters, U.S. Air Force, November 4, 2016.

Annex 3-01, *Counterair Operations*, Washington, D.C.: Headquarters, U.S. Air Force, October 27, 2015.

Annex 3-03, *Counterland Operations*, Washington, D.C.: Headquarters, U.S. Air Force, March 17, 2017.

Annex 3-05, *Special Operations*, Washington, D.C.: Headquarters, U.S. Air Force, February 9, 2017.

Annex 3-10, *Force Protection*, Washington, D.C.: Headquarters, U.S. Air Force, April 17, 2017.

Annex 3-12, *Cyberspace Operations*, Washington, D.C.: Headquarters, U.S. Air Force, November 30, 2011.

Annex 3-13, *Information Operations*, Washington, D.C.: Headquarters, U.S. Air Force, April 28, 2016.

Annex 3-17, *Air Mobility Operations*, Washington, D.C.: Headquarters, U.S. Air Force, April 5, 2016.

Annex 3-2, *Irregular Warfare*, Washington, D.C.: Headquarters, U.S. Air Force, July 12, 2016.

Annex 3-22, *Foreign Internal Defense*, Washington, D.C.: Headquarters, U.S. Air Force, July 2015.

Annex 3-27, *Homeland Operations*, Washington, D.C.: Headquarters, U.S. Air Force, April 28, 2016.

Annex 3-34, *Engineer Operations*, Washington, D.C.: Headquarters, U.S. Air Force, December 30, 2014.

Annex 3-51, *Electronic Warfare*, Washington, D.C.: Headquarters, U.S. Air Force, October 10, 2014.

Annex 3-52, *Airspace Control*, Washington, D.C.: Headquarters, U.S. Air Force, July 21, 2014.

Annex 3-60, *Targeting*, Washington, D.C.: Headquarters, U.S. Air Force, February 14, 2017.

Annex 3-61, *Public Affairs Operations*, Washington, D.C.: Headquarters, U.S. Air Force, June 19, 2014.

Annex 3-70, *Strategic Attack*, Washington, D.C.: Headquarters, U.S. Air Force, May 25, 2017.

Annex 3-72, *Nuclear Options*, Washington, D.C.: Headquarters, U.S. Air Force, May 19, 2015.

Annex 4-0, *Combat Support*, Washington, D.C.: Headquarters, U.S. Air Force, December 21, 2015.

Annex 4-02, *Medical Operations*, Washington, D.C.: Headquarters, U.S. Air Force, September 29, 2015.

ASPJ—See *Air & Space Power Journal.*

Assistant Secretary of the Air Force (Financial Management and Comptroller), *United States Air Force Statistical Digest, Fiscal Year 1996*, Washington, D.C.: Headquarters, U.S. Air Force, June 1997.

———, *United States Air Force Statistical Digest, Fiscal Year 2000*, Washington, D.C.: Headquarters, U.S. Air Force, 2001.

———, *United States Air Force Statistical Digest, Fiscal Year 2004*, Washington, D.C.: Headquarters, U.S. Air Force, 2005.

———, *United States Air Force Statistical Digest, Fiscal Year 2010*, Washington, D.C.: Headquarters, U.S. Air Force, 2011.

Avant, Deborah D., *Political Institutions and Military Change: Lessons from Peripheral Wars*, Ithaca, N.Y.: Cornell University Press, 1994.

Berg, Paul D., "Vignettes, Doctrine NOTAMs, and the Latest *Chronicles* Articles," *Air & Space Power Journal*, Vol. 19, No. 1, Spring 2005, p. 17.

Boudreau, Robert N., "The New AFM 1-1: Shortfall in Doctrine?" *Airpower Journal*, Vol. 6, No. 4, Winter, 1992, pp. 37–45.

Builder, Carl H., *The Masks of War: American Military Styles in Strategy and Analysis*, Baltimore, Md.: Johns Hopkins University Press, 1989.

———, *The Icarus Syndrome: The Role of Air Power Theory in the Evolution and Fate of the U.S. Air Force*, New Brunswick, Conn.: Transaction Publishers, 1994.

Chairman of the Joint Chiefs of Staff Instruction 1800.01E, *Officer Professional Military Education Policy*, Washington, D.C.: Joint Chiefs of Staff, May 29, 2015.

Chairman of the Joint Chiefs of Staff Instruction 5120.02D, *Joint Doctrine Development System*, Washington, D.C.: Joint Chiefs of Staff, January 5, 2015.

Chairman of the Joint Chiefs of Staff Manual 5120.01A, *Joint Doctrine Development Process*, Washington, D.C.: Joint Chiefs of Staff, December 29, 2014.

CJCSI—*See* Chairman of the Joint Chiefs of Staff Instruction.

Curry, Bruce L., *Turn Points in the Air: Using Historical Examples to Illustrate USAF Doctrine*, Maxwell Air Force Base, Ala.: Air War College, Air University, 1997.

Curtis E. LeMay Center for Doctrine Development and Education, homepage, undated. As of October 31, 2017:
http://www.doctrine.af.mil/

———, *Core Doctrine,* Vol. I, *Basic Doctrine*, Washington, D.C.: Headquarters, U.S. Air Force, February 27, 2015a. As of June 4, 2018:
http://www.doctrine.af.mil/Core-Doctrine/Vol-1-Basic-Doctrine/

———, *Core Doctrine,* Vol. II, *Leadership*, Washington, D.C.: Headquarters, U.S. Air Force, August 8, 2015b. As of June 4, 2018:
http://www.doctrine.af.mil/Core-Doctrine/Vol-2-Leadership/

———, *Core Doctrine,* Vol. III, *Command*, Washington, D.C.: Headquarters, U.S. Air Force, November 22, 2016. As of June 4, 2018:
http://www.doctrine.af.mil/Core-Doctrine/Vol-3-Command/

———, "Intermediate Courses," webpage, Air University, April 27, 2017a. As of November 10, 2017:
http://www.airuniversity.af.mil/LeMay/Display/Article/1099686/intermediate-courses/

———, "Senior Leader Courses," webpage, Air University, April 27, 2017b. As of November 10, 2017:
http://www.airuniversity.af.mil/LeMay/Display/Article/1099524/senior-leader-courses/

———, in-person, phone, and email discussions with RAND project team, Maxwell Air Force Base, Ala., July–November 2017c.

Department of Defense Directive No. 5100.01, *Functions of the Department of Defense and Its Major Components*, Washington, D.C., U.S. Department of Defense, December 21, 2010.

Dick, Amanda, "Commander Sets Priorities, Way Ahead for 9th AF," *News: 9th Air Force,* February 2, 2017. As of November 9, 2017:
http://www.9af.acc.af.mil/News/Article/1068342/commander-sets-priorities-way-ahead-for-9th-af/

DoD—*See* U.S. Department of Defense.

DoD Directive—*See* Department of Defense Directive.

Drew, Dennis M., "Inventing a Doctrine Process," *Airpower Journal*, Vol. 9, No. 4, Winter 1995, pp. 42–52.

Dunlap, Charles J., Jr., "Air-Minded Considerations for Joint Counterinsurgency Doctrine," *Air & Space Power Journal*, Vol. 21, No. 4, Winter 2007, pp. 63–74.

Ehrhard, Thomas P., *An Air Force Strategy for the Long Haul*, Washington, D.C.: Center for Strategic and Budgetary Assessments, 2009.

Ehrhart, Robert C., "Some Thoughts on Air Force Doctrine," *Air University Review*, March–April 1980.

Fernandez, Sergio, and Hal G. Rainey, "Managing Successful Organizational Change in the Public Sector," *Public Administration Review,* Vol. 66, No. 2, March 2006, pp. 168–176.

Finney, Robert T., *History of the Air Corps Tactical School, 1920–1940*, Maxwell Air Force Base, Ala.: Research Studies Institute, U.S. Air Force, Historical Division, Air University, 1955.

Fogleman, Ronald R., "Aerospace Doctrine: More Than Just a Theory," *Airpower Journal*, Vol. 10, No. 2, 1996, pp. 40–47.

Furr, William F., "Joint Doctrine: Progress, Prospects, and Problems," *Airpower Journal*, Vol. 5, No. 3, Fall 1991, pp. 36–45.

Futrell, Robert Frank, "Chapter 3: The Air Force in a Changing Defense Environment," in *Ideas, Concepts, Doctrine: Basic Thinking in the United States Air Force, 1961–1984*, Vol. II, Maxwell Air Force Base, Ala.: Air University Press, 1989a.

———, "Chapter 7: The Air Force Writes Its Doctrine, 1947–55," in *Ideas, Concepts, Doctrine: Basic Thinking in the United States Air Force, 1907–1960*, Vol. I, Maxwell Air Force Base, Ala.: Air University Press, 1989b.

Goldfein, Dave, *CSAF Focus Area: Strengthening Joint Leaders and Teams*, Washington, D.C.: U.S. Air Force, 2016.

Grant, Rebecca, "Closing the Doctrine Gap," *AIR FORCE Magazine*, January 1997, pp. 48–52.

Green, Thomas H., *The Development of Air Doctrine in the Army Air Arm, 1917–1941*, Washington, D.C.: Office of Air Force History, 1985.

Grissom, Adam, "The Future of Military Innovation Studies," *Journal of Strategic Studies*, Vol. 29, No. 5, 2006, pp. 905–934. As of April 23, 2018: http://dx.doi.org/10.1080/01402390600901067

Hallion, Richard P., "Doctrine, Technology, and Air Warfare: A Late Twentieth-Century Perspective," *Airpower Journal*, Vol. 1, No. 2, Fall 1987, pp. 16–27.

Harrell, Margaret C., and Melissa A. Bradley, *Data Collection Methods: Semi-Structured Interviews and Focus Groups*, Santa Monica, Calif.: RAND Corporation, TR-718-USG, 2009. As of April 16, 2018: https://www.rand.org/pubs/technical_reports/TR718.html

Harvard, James W., "Airmen and Mission Command," *Air & Space Power Journal*, March–April 2013, pp. 131–146.

Headquarters, Department of the Army, *Mission Command*, Washington, D.C., ADP 6-0, May 2012, incorporating changes as of March 12, 2014a.

———, *Doctrine Primer*, Washington, D.C., ADP 1-01, September 2014b. As of April 23, 2018: http://armypubs.army.mil/epubs/DR_pubs/DR_a/pdf/web/adp1_01.pdf

Headquarters, U.S. Air Force, *Point Paper on the Functions of New Air Force Doctrine Center*, Maxwell Air Force Base, Ala., 1996.

———, *Air Force Glossary*, Washington, D.C., 2016.

Høiback, Harald, "What Is Doctrine?" *Journal of Strategic Studies*, Vol. 34, No. 6, December 1, 2011, pp. 879–900. As of April 23, 2018: http://dx.doi.org/10.1080/01402390.2011.561104

———, "The Anatomy of Doctrine and Ways to Keep It Fit," *Journal of Strategic Studies*, Vol. 39, No. 2, 2016, pp. 185–197. As of April 23, 2018: http://dx.doi.org/10.1080/01402390.2015.1115037

Holley, I. B., Jr., "A Modest Proposal: Making Doctrine More Memorable," *Airpower Journal*, Vol. 9, No. 4, 1995, pp. 14–20.

Hukill, Jeffrey, Larry Carter, Scott Johnson, Jennifer Lizzol, Edward Redman, and Panayotis Yannakogeorgos, *Air Force Command and Control: The Need for Increased Adaptability*, Maxwell Air Force Base, Ala.: Air Force Research Institute, Air Force Research Institute Papers 2012-5, July 2012.

Hutchens, Michael E., William D. Dries, Jason C. Perdew, Vincent D. Bryant, and Kerry E. Moores, "Joint Concept for Access and Maneuver in the Global Commons: A New Joint Operational Concept," *Joint Forces Quarterly*, Vol. 84, First Quarter 2017, pp. 134–139.

J-7—*See* Joint Force Development Directorate of the Joint Staff.

Johnson, David E., *Learning Large Lessons: The Evolving Roles of Ground Power and Air Power in the Post–Cold War Era*, Santa Monica, Calif.: RAND Corporation, MG-405-1-AF, 2007. As of April 23, 2018: https://www.rand.org/pubs/monographs/MG405-1.html

Johnston, Paul, "Doctrine Is Not Enough: The Effect of Doctrine on the Behavior of Armies," *Parameters*, Autumn 2000, pp. 30–39. As of April 23, 2018: http://ssi.armywarcollege.edu/pubs/parameters/articles/00autumn/johnston.htm

Joint Chiefs of Staff, *Capstone Concept for Joint Operations: Joint Force 2020*, Washington, D.C.: U.S. Department of Defense, September 10, 2012. As of April 18, 2018: http://www.defenseinnovationmarketplace.mil/resources/JV2020_Capstone.pdf

Joint Force Development Directorate of the Joint Staff, in-person and email discussions with RAND project team, Pentagon, Washington, D.C., July–October 2017.

Joint Publication 1, *Doctrine of the Armed Forces of the United States*, Washington, D.C.: Joint Chiefs of Staff, March 25, 2013.

Joint Publication 2-0, *Joint Intelligence*, Washington, D.C.: Joint Chiefs of Staff, October 22, 2013.

Joint Publication 2-01, *Joint and National Intelligence Support to Military Operations*, Washington, D.C.: Joint Chiefs of Staff, January 5, 2012.

Joint Publication 2-01.3, *Joint Intelligence Preparation of the Operational Environment*, Washington, D.C.: Joint Chiefs of Staff, May 21, 2014.

Joint Publication 2-03, *Geospatial Intelligence in Joint Operations*, Washington, D.C.: Joint Chiefs of Staff, October 31, 2012.

Joint Publication 3-0, *Joint Operations*, Washington, D.C.: Joint Chiefs of Staff, January 17, 2017.

Joint Publication 3-01, *Countering Air and Missile Threats*, Washington, D.C.: Joint Chiefs of Staff, April 21, 2017.

Joint Publication 3-03, *Joint Interdiction*, Washington, D.C.: Joint Chiefs of Staff, September 9, 2016.

Joint Publication 3-05, *Special Operations*, Washington, D.C.: Joint Chiefs of Staff, July 16, 2014.

Joint Publication 3-12 (R), *Cyberspace Operations*, Washington, D.C.: Joint Chiefs of Staff, February 5, 2013.

Joint Publication 3-13, *Information Operations*, Washington, D.C.: Joint Chiefs of Staff, November 2012, incorporating Change 1, November 2014.

Joint Publication 3-22, *Foreign Internal Defense*, Washington, D.C.: Joint Chiefs of Staff, July 12, 2010.

Joint Publication 3-27, *Homeland Defense*, Washington, D.C.: Joint Chiefs of Staff, July 29, 2013.

Joint Publication 3-28, *Defense Support of Civil Authorities*, Washington, D.C.: Joint Chiefs of Staff, July 31, 2013.

Joint Publication 3-30, *Command and Control of Joint Air Operations*, Washington, D.C.: Joint Chiefs of Staff, February 10, 2014.

Joint Publication 3-34, *Joint Engineer Operations*, Washington, D.C.: Joint Chiefs of Staff, January 6, 2016.

Joint Publication 3-52, *Joint Airspace Control*, Washington, D.C.: Joint Chiefs of Staff, November 13, 2014.

Joint Publication 3-59, *Meteorological and Oceanographic Operations*, Washington, D.C.: Joint Chiefs of Staff, December 7, 2012.

Joint Publication 3-60, *Joint Targeting*, Washington, D.C.: Joint Chiefs of Staff, January 31, 2013.

Joint Publication 4-0, *Joint Logistics*, Washington, D.C.: Joint Chiefs of Staff, October 16, 2013.

Joint Publication 5-0, *Joint Operation Planning*, Washington, D.C.: Joint Chiefs of Staff, August 11, 2011.

JP—*See* Joint Publication.

Kronvall, Olof, and Magnus Petersson, "Doctrine and Defence Transformation in Norway and Sweden," *Journal of Strategic Studies*, Vol. 39, No. 2, February 23, 2016, pp. 280–296. As of April 23, 2018:
http://dx.doi.org/10.1080/01402390.2015.1115040

Lambeth, Benjamin S., *The Unseen War: Allied Air Power and the Takedown of Saddam Hussein*, Annapolis, Md.: Naval Institute Press, 2013.

Lambright, W. Henry, "Leadership and Change at NASA: Sean O'Keefe as Administrator," *Public Administration Review*, Vol. 68, No. 2, March–April 2008, pp. 230–240.

Lee, Caitlin, Bart E. Bennett, Lisa M. Harrington, and Darrell D. Jones, *Rare Birds: Understanding and Addressing Air Force Underrepresentation in Senior Joint Positions in the Post–Goldwater Nichols Era*, Santa Monica, Calif.: RAND Corporation, RR-2089-AF, 2017. As of April 23, 2018:
https://www.rand.org/pubs/research_reports/RR2089.html

LeMay Center—*See* Curtis E. Lemay Center for Doctrine Development and Education.

Libicki, Martin C., *Conquest in Cyberspace: National Security and Information Warfare*, New York: Cambridge University Press, 2007.

Long, Austin, *The Soul of Armies: Counterinsurgency Doctrine and Military Culture in the US and UK*, Ithaca, N.Y.: Cornell University Press, 2016.

McLean, Brian, "Reshaping Centralized Control/Decentralized Execution for the Emerging Operating Environment," *Over the Horizon*, March 13, 2017. As of April 23, 2018:
https://overthehorizonmdos.com/2017/03/13/reshaping-cc-dc-de/

Meilinger, Phillip S., "The Problem with Our Airpower Doctrine," *Airpower Journal*, Spring 1992.

Mowbray, James A., "Air Force Doctrine Problems: 1926–Present," *Air & Space Power Journal*, Winter 1995, pp. 1–17.

Nellis Air Force Base, "Exercises and Flight Operations," webpage, undated. As of February 5, 2018:
http://www.nellis.af.mil/Home/Flying-Operations/

———, "561st Joint Tactics Squadron," fact sheet, March 24, 2016. As of November 10, 2017: http://www.nellis.af.mil/About/Fact-Sheets/Display/Article/703580/561st-joint-tactics-squadron/

Posen, Barry R., *The Sources of Military Doctrine: France, Britain, and Germany Between the World Wars*, Ithaca, N.Y.: Cornell University Press, 1984.

———, "Military Doctrine and the Management of Uncertainty," *Journal of Strategic Studies*, Vol. 39, No. 2, 2016, pp. 159–173.

Powell, Colin L., "A Word from the Chairman," *Joint Force Quarterly*, Vol. 1, No. 1, 1993, p. 5.

Romjue, John L., *American Army Doctrine for the Post-Cold War*, TRADOC Historical Monograph Series, Fort Monroe, Va.: Military History Office, U.S. Army Training and Doctrine Command, 1997.

Rosen, Stephen Peter, *Winning the Next War: Innovation and the Modern Military*, Ithaca, N.Y.: Cornell University Press, 1994.

Snyder, Jack, *The Ideology of the Offensive: Military Decision Making and the Disasters of 1914*, Ithaca, N.Y.: Cornell University Press, 1989.

Spirtas, Michael, Thomas Young, and S. Rebecca Zimmerman, *What It Takes: Air Force Command of Joint Operations*, Santa Monica, Calif.: RAND Corporation, MG-777-AF, 2009. As of April 17, 2018:
https://www.rand.org/pubs/monographs/MG777.html

U.S. Air Force, *Air Force Future Operating Concept: A View of the Air Force in 2035*, Washington, D.C., 2015. As of April 20, 2018:
http://www.ang.af.mil/Portals/77/documents/AFD-151207-019.pdf

U.S. Air Force Expeditionary Center, "423d Mobility Training Squadron," August 20, 2015. As of November 3, 2017:
http://www.expeditionarycenter.af.mil/About-Us/Fact-Sheets/Display/Article/787892/423d-mobility-training-squadron/

U.S. Department of Defense, *Joint Operational Access Concept (JOAC)*, Washington, D.C., January 17, 2012.

———, *DoD Dictionary of Military and Associated Terms*, Washington, D.C., March 2017.

Wilson, Heather, "State of the Air Force," presentation at 2017 Air Force Association Air, Space & Cyber Conference, National Harbor, Md., September 18, 2017.

Zisk, Kimberly Marten, *Engaging the Enemy: Organization Theory and Soviet Military Innovation, 1955–1991*, Princeton, N.J.: Princeton University Press, 1993.